HOW TO MAKE A WETLAND

HOW TO MAKE A WETLAND

Water and Moral Ecology in Turkey

Caterina Scaramelli

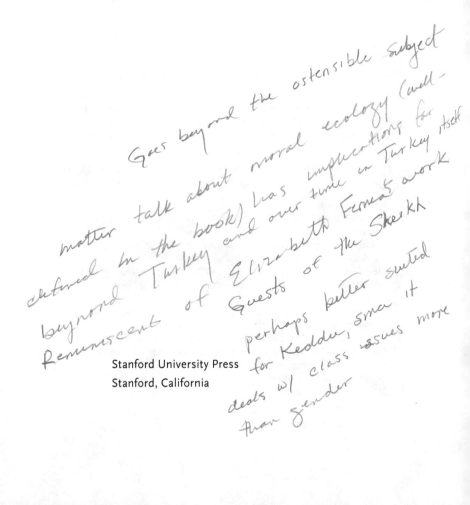

Goes beyond the ostensible subject

matter talk about moral ecology (well-

defined in the book) has implications far

beyond Turkey and our time in Turkey itself

Reminiscent of Elizabeth Ferea's work

Guests of the Sheikh

perhaps better suited

for Keddie, since it

deals w/ class issues more

than gender

Stanford University Press
Stanford, California

Stanford University Press
Stanford, California

Printed in the United States of America on acid-free, archival-quality paper

Library of Congress Cataloging-in-Publication Data

Names: Scaramelli, Caterina, author.
Title: How to make a wetland : water and moral ecology in Turkey / Caterina Scaramelli.
Description: Stanford, California : Stanford University Press, 2021. | Includes bibliographical references and index.
Identifiers: LCCN 2020027650 (print) | LCCN 2020027651 (ebook) | ISBN 9781503613850 (cloth) | ISBN 9781503615403 (paperback) | ISBN 9781503615410 (ebook)
Subjects: LCSH: Wetland conservation—Social aspects—Turkey. | Wetland conservation—Moral and ethical aspects—Turkey. | Wetland conservation—Political aspects—Turkey. | Human ecology—Turkey.
Classification: LCC QH77.T9 S33 2021 (print) | LCC QH77.T9 (ebook) | DDC 333.91/809561—dc23
LC record available at https://lccn.loc.gov/2020027650
LC ebook record available at https://lccn.loc.gov/2020027651

Cover art: Acrylic painting by Lokman Önsoy
Cover design: Rob Ehle

Contents

Maps and Illustrations

Acknowledgments

I am immensely grateful to my Turkish friends, hosts, and interlocutors who have made this work possible. My biggest thanks go to people of the Kızılırmak and Gediz Deltas. This book is just a small fragment of all the things I learned from you. Special thanks go to many hosts and friends and to the wetland experts who patiently answered my questions: Ahmet, Alaattin and Cemile, Ali Kemal *hoca*, Arzu *hoca*, Avni and Hatice, Bilgi and Sunay, Burak, Burçin, Burcu, Elif *hoca*, Emrah, Engin, Ercan *hoca*, Erdal, Esra, Gazanfer and Nuray, Güven, Hatice *hoca*, Hülya, Kadir and Melek, Kiraz, Kübra, Lokman and Gülşah, Oral, Özlem A., Özlem D., Ömer and Özlem, Ömer Faruk, Piero, Seçil, Sevim and Turgut, Tulay and Ahmet, Veli, and Yücel and Güldan. Faculty at Ondokuz Mayıs University, Ege University, and the Döküz Eykül University; Nature Conservation and National Parks staff; countless government and city officials; staff at Doğa Derneği; the Tour du Valat team; and the Çamaltı Saltworks personnel were very gracious and patient with my questions. Members of KoşAnkara and the Ege University climbing club took me from the swamps back to the mountains. During fieldwork, I was amazed by the effervescent kindness and wisdom of Giulia Tacchini and Kalliopi Amygdalou, who have now become family, with Marco, Kostas, and Annamaria. My beloved friends Aykut Öztürk, Ced Öner, and Ceylan Dökmen hosted me more times than I can count. My year at Boğaziçi University was life changing. I am grateful to have learned with Ayfer Bartu Candan and Koray Çalışkan and to have worked with John Scott, Berrin Torolosan, and

Michael Hornsby. Since those times, Marlene Schäfers, now a fellow anthropologist, has been a good friend to think with.

At MIT, I was fortunate to find such brilliant and caring scholars, who have all left a deep imprint on my work. Christine Walley, whose writings inspired me to become an anthropologist, taught me the anthropology of science, environment, and documentary film with much clarity, care, and empathy. Harriet Ritvo introduced me to environmental history and to thinking with animals. Lerna Ekmekçioğlu inspired me with her sharp writing and thinking, pushing me to better attend to the multiple histories of communities in the wetland. I have endless gratitude for the many things I learned from Stefan Helmreich—about fieldwork, science, water, theory—and also how to be a kind scholar. Stefan offered consistent support in all matters of research and writing and read hundreds of drafts, all marked with incisive questions and groundbreaking suggestions. I was also lucky to work with Chris Boebel, Manduhai Buyandelger, Michael Fisher, David Jones, David Keiser, Clapperton Chakanetsa Mavhunga, Heather Paxson, Hannah Rose Shell, and Susan Silbey. Fellow researchers offered companionship and feedback on different parts of this project, especially Luisa Castro, Ashawari Chaudhuri, Nadia Christidi, Lisa Messeri, Peter Oviatt, Shira Shmuely, Michaela Thompson, Emily Wanderer, and Rebecca Woods. Participants in the Harvard Anthropology's Political Ecology Working Group offered incisive critiques and constructive feedback. Special thanks go to Steve Caton, Ekin Kurtiç, Ajantha Subramanian, Rachel Thompson, and Dilan Yıldırım for animating the group.

At Amherst College, I found a nurturing intellectual home at the Center for Humanistic Inquiry and in the Department of Anthropology; I thank the Keiter Fellowship and colleagues and staff for their support during these magical two years. Big thanks go to the friends and colleagues at Amherst who read and commented on parts of this book: Brett Brehm, Li Cornfeld, Chris Dole, Fred Errington, Deborah Gewertz, Reed Gochberg, Amy C. Hall, Hannah Holleman, Colleen Kim Daniher, Amy Johnson, Ada Link, Jennifer Pranolo, Maria Sidorkina, and Martha Umphrey. Abbas Shah helped with his incisive student perspective on my writing and with bibliographic research. I completed revisions while at Boston University, and I am thankful for the support of faculty members and staff at in the Departments of Anthropology and of Earth and Environment. Kimberly Arkin, Joanna Davidson, Fallou Ngom, Parker Shipton, and Robert Weller asked productive questions on parts of this work. I am grateful to my chairs, Guido Salvucci and Nancy Smith-Hefner, and to my dean, Nazli Kibria. Scholars in Jeff Rubin's Seeing and Not Seeing (SANS) seminar and in Anne Short-Gianotti's Human-Environment working group offered much-needed feedback and inspiration. For their detailed comments on chapter drafts, I thank Sultan Doughan, Calynn Dower, Emile Edelblutte, Ayşe Parla, Merav Shohet, Anne Short-Gianotti, and Kira Sullivan-Wiley.

The Social Science Research Council allowed me to develop this project when it was a dormant seed of an idea, and our interdisciplinary cohort of "ecological history" fellows, Paolo Bocci Angelo Cagliotti, Zach Caple, Sam Dolbee, Nate Ela, David Fedman, Jenny Goldstein, Tim Johnson, Laura Martin, Maria Taylor, and Greg Thaler, grew into a steady community of colleagues and co-conspirators. Peter Perdue and Steve Harrell got us started and kept us going over the years. MIT, the National Science Foundation, and the Wenner-Gren Foundation for Anthropological Research funded fieldwork in Turkey. I completed this book thanks to a wonderful year at the Agrarian Studies program at Yale's MacMillan Center. James Scott and Kalyanakrishnan Shivaramakrishnan offered generous comments and constructive critiques. I could not have wished for a better group of fellow agrarians to share drafts, cannolis, and animated discussions over copious coffee: Tony Andersson, John Buchanan, Sakura Christmas, Kathryn De Luna, and Keri Lambert turned the shared office into an inspiring writing cove. I have benefited tremendously from presenting parts of this work at Amherst College, Boston University, Harvard University, the London School of Economics, MIT, Tallin University, the University of Chicago, and the University of Vienna, as well as at the annual meetings of the American Anthropological Association, the Middle East Studies Association, the European Association of Social Anthropologists, the Canadian Anthropology Society, the Association of Social Anthropologists of the UK and Commonwealth, and the Society for the Social Studies of Science. For offering insights and critiques, I am grateful to Cemil Aksu, Oğuz Alyanak, Nikhil Anand, Delal Aydın, Elif Babül, Les Beldo, Filippo Bertoni, Ashley Carse, Timothy Choy, Luisa Cortesi, Can Dalyan, Sinan Erensu, Hatice Erten, Can Evren, Erdem Evren, Bilge Fırat, Stefanie Graeter, Colin Hoag, Hazal Hurman, Onur Inal, Sheila Jasanoff, Eban Kirksey, Yavuz Köse, James McCann, Alan Mikhail, Tamar Novick, Yağmur Nurhat, Zeynep Oğuz, Mathijs Pelkmans, Tanya Richardson, Seda Saluk, Hande Sarıkuzu, Brian Silverstein, Rebecca Zarger, Sezai Ozan Zeybek, and Amy Zhang. At Stanford University Press, Kate Wahl, Caroline McKusick, Tim Roberts, Cynthia Lindlof, Rob Ehle, Stephanie Adams, and the rest of the editorial team took fantastic care of this book. I also wish to thank the two anonymous peer reviewers.

I could not write without the steady and loving presence of my writing group on Middle East environments: Jessica Barnes, Tessa Farmer, Simone Popperl, Kali Rubaii, and Sophia Stamatopoulou-Robbins have read and commented on my work from the very early stages to its completion. My debt to them is endless. My brilliant friends Nicole Labruto and Alessandro Angelini have been a constant source of inspiration, wonder, and adventure. Besides offering essential feedback on many drafts, brewing a lot of coffee, and following me on rusty village bicycles in the delta, Mary Kuhn and David Singerman indulged in many long conversations about this research, provided life-changing advice, and continue to be the best companions for environmental forays, open-ended conversations, and more. Many friends and fellow anthropologists offered comments, critiques, feedback, and conversations that shaped my thinking on the politics and poetics of environment. I am indebted to Naor Ben-Yehoyada, Yael Berda, Asia Bojczewska, Anna Calia, Namita Dharia, Richard Delacy, Nejat Dinç, Diana Doty, Tasha Eccles, Morgan Frank, Ludovica Gazze, Radikha Govindrajan, Jeff Kahn, Hayden Kantor, Roanne Kantor, Franz Krause, Ekin Kurtiç, Jyothi Natarajan, Jen and Greg Noble, Adi Nochur, Liza Oliver, Mariana Pote, Casey Riley, Alicia Ringel, Joshua Specht, Chiara Superti, Anand Vaidya, Joanna Ware, and Aslı Zengin.

I am fortunate to be surrounded by my wonderful and loving extended family, in-laws, and chosen family, from Michigan to Lombardy. Many hosts and research interlocutors in Turkey have become like family. I probably would not have embarked on this work if my parents had not taken me with them to Turkey as a toddler and many times since. My father, Daniele Scaramelli, instilled within me his curiosity, sense of humor, and an open mind—he did not hesitate to step barefoot into a wetland lake and to join me on many fieldwork excursions. I wish he were still with us. My mother, Maria Corno, who traveled on foot from her home in Milan to visit me in Turkey, showed me the value of slowing down, gave me quiet places to write when I needed them, and continues to be a source of encouragement and wisdom.

Benjamin Siegel has read countless drafts, accompanied me on research trips, and taught me that (almost) nothing is too big to be tackled. Ben's wisdom with words, skillful ingenuity, adventurous playfulness, superb cooking, unconditional generosity, and expansive love sustained me during years of research and writing. Here's to many more adventures!

HOW TO MAKE A WETLAND

Introduction

The Watery Place

On a very hot day in August 2012, at around midday, I got stuck. My feet were sucked deeper in the mud and, trying to lift them up, I fell face down into the muck of the marsh. A few meters away, my friend Deniz, a marine biologist, was knee-deep in water, struggling to make another step.[1] Emre, an ornithologist, had donned fisherman-style plastic overalls that went up to his chest, and he kept walking swiftly, holding a GPS device in his outstretched arm, soon disappearing behind a tall thicket of reeds. I wiggled out of my orange plastic boots, which I had bought the day before from a fishing-supply store at Deniz's insistence. I stood up in the hot breeze. Deniz and I continued walking barefoot in the soft mud. We held our boots and soggy socks in one hand, and we started to run to catch up with Emre.

At that moment, I did not understand the importance of the claim Emre was staking as he geolocated the wetland reeds in the marsh, using his GPS to record the coordinate points of the reeds' meandering in the marsh. Later, I would learn that the reeds were one of many nonhuman actors in a political drama unfolding in a rapidly changing delta, for the reeds marked a boundary between different water salinities. Encroached upon by fields of water-thirsty cash crops, industrial areas, and new exurbs, the wetland had become an endangered ecological refuge, a place where Turkish scientists, bureaucrats, and activists *made* nature. The reeds were watered by the same irrigation infrastructure that diverted riverine flows away from the coastal marshes.

Figure 1. Mapping the reeds in the Gediz Delta salt marshes, 2013. Photo by the author.

Conservation legislation and environmental governance had allowed the wetland continue to exist, though it had displaced fishers and farmers from marshes and swamps—a form of everyday environmental violence that took shape atop the sediments of a longer history of ethnic violence, dispossession, and population resettlement.

The paradoxical fate of the marsh—at once an ecological refuge and a site of violence and conflict—was entangled with other socio-ecological transformations that have come to hold moral significance for contemporary Turkish environmental scientists, fishermen, farmers, bureaucrats, planners, and nongovernmental organization (NGO) workers. While Deniz, Emre, and I walked among the reeds, Turkish environmental NGOs were campaigning to draw attention to the rapid disappearance of wetlands nationwide. These watery biomes were dredged for agricultural and industrial development, polluted with urban and industrial wastewater, and flooded in the basins of dam reservoirs. Small-scale farmers and groups of urban middle-class residents had mobilized on local and national water rights platforms, crafting shared struggles against the expansion of hydropower infrastructure across thousands of rivers throughout Turkey. Civil-society groups had called attention to radioactive waste buried in city fields. Environmental scientists had published reports of waterways and soil polluted with carcinogenic substances.

The surface water of the marsh was brown, yellow, and light gray, and it was scalding hot; as I walked, my toes wiggled in darker, thicker, cooler mud. Each step released an acrid and sulfurous smell, tempered by the lighter scent of salt, tree bark, and pungent grass. I could hear the rustling of reeds, a polyphony of birdcalls, and the splashing and suction of our steps.

"What do you like most about this landscape?" I asked Emre, as he handed me the GPS device to take over the reed mapping.

"Are you asking so that you can write this in your thesis?" he asked, with a smile.

I pleaded guilty as charged. "Why is the mud changing color?" I continued, insistent.

"This is the work of anoxic bacteria," he said.

Deniz took a handful of mud and let it slip through his fingers. "This is a nourishing soup for biodiversity. For example, the mighty *Artemia salina* [a brine shrimp] lives here and eats microscopic algae. And then the flamingos eat *Artemia*. It's all connected. And it all comes together in the wetland." Deniz used a Turkish neologism for wetland, *sulakalan*, meaning "watery place."

Emre, Deniz, and I stepped on drier land and sat down on the cracked mud, sharing cookies and water. We were in the lower Gediz Delta, in the northern shore of Izmir Bay, on Turkey's Aegean Sea, walking across what had once been a saltpan. A conservation official had dropped us off on one side of the saltpan hours earlier and would be waiting for us on the other side. Only wetland conservation management and the hunt-control patrols were allowed motorized access to the wetland conservation area. In the early 1980s, the Çamaltı Saltworks had started expanding on the adjacent coastal marshes. Concerned natural scientists, birders, hunters, and environmental activists had organized to stop this infrastructural unmaking of the marshes. This effort had led to the creation of a nature conservation area, which people often called Kuş Cenneti, the Bird Paradise.

The lower Gediz Delta was also one of Turkey's fourteen wetlands that appeared in the Ramsar Convention list of Wetlands of International

Importance, and it held other overlapping national conservation statuses. In 2017, scientists at a Turkish environmental NGO along with university scientists from natural science departments would jointly write a report demonstrating that the delta fit the criteria to become a UNESCO World Heritage Natural Site: to be of "outstanding" value and "to contain the most important and significant natural habitats for in-situ conservation of biological diversity."[2] The same year, environmental advocates in Izmir organized around the call "Don't touch my flamingo" and filed a civil lawsuit against the Ministry of Transportation to prevent a highway bridge from being constructed on the wetland.[3] In 2019, a Turkish NGO orchestrated a social media campaign called "Gediz is our heritage," collecting signatures in support of the delta's UNESCO candidacy.[4]

The Gediz Delta wetland was not only a site of environmental advocacy: its flowing materiality was entangled with other overlapping infrastructural transformations of the delta's agrarian and industrial landscape. Saltworks infrastructure—pans, dikes, roads, and canals—remained in place but had now become part of the "natural" Gediz wetlands conservation area. The active sections of the saltworks now hosted artificial islands for flamingos. The saltpans had become a popular destination for scientists, nature photographers, and local tourists, who flocked there to observe thousands of flamingos feeding in the *Artemia*-rich water and thousands of birds from hundreds of species. From the salt marshes rose the sparsely forested hills of Üç Tepeler, once the site of the ancient fortified city of Leucae.[5]

Fishers on small motorized dinghies cast their nets in the shallow waters of the lagoons and the shores of the Aegean Sea. Upstream in the Gediz Delta, farmers and migrant day laborers worked to reap seasonal harvests of cotton, grains, and vegetables. Adjacent to an industrial leather district and a new university campus built among the cotton fields was one of Turkey's biggest villa and apartment complexes, where opportunistic real estate speculators enticed members of Izmir's aspiring (and declining) middle classes who could no longer afford a city apartment. Izmir's wastewater plant collected the city's sewage and then pumped treated water into the sea near the industrial area. A eucalyptus orchard, planted on drained coastal marshes, gave way to an experimental forest, a designated picnic area, and then a zoological park. A long bike path connected Izmir's wealthy neighborhood of Bostanlı to the wetland conservation area's headquarters, where Emre, the biologist, worked.

This book is about how environments saturated in water—such as this deltaic salt marsh—*became* wetlands, a transformation that has been at once political, cultural, and material. During my ethnographic fieldwork research, between 2012 and 2018, I followed the making of Turkish wetlands in two deltaic environments. I moved between the Gediz Delta, on the Aegean Sea, and the Kızılırmak Delta, on the Black Sea coast of Turkey. Drawing comparisons between different wetland ecologies is central to how environmental scientists, state officials, and NGOs make Turkish wetlands into objects of knowledge and management.[6] ← problematic (vague) footnote

One day in October 2013, I followed Emre as he caught a flight from Izmir to Samsun and then drove to the Kızılırmak Delta with his manager and office staff. I sat in a dark conference room as Emre gave a talk to a small audience of scientists, environmental managers, and provincial administrators in the management headquarters of the Kızılırmak Delta wetlands. Behind us, farmers on horseback and on small boats directed their herds of water buffaloes as they swam across a wetland lake, returning to the farms after six months of grazing in the marshes. Flocks of birds assembled into V-shaped formations before setting out on their migratory journey southward.

Emre described the management system that governed the Gediz Wetlands—an institutional form constituted by a partnership of delta municipalities and overseen by Izmir's provincial government. The National Parks Bureau contracted this organization for implementing its wetland management guidelines on-site. Two years later, an identical administrative structure was implemented in the Kızılırmak Delta. At that later time, I was living in a local village with a family of buffalo and rice farmers. The delta's farmers, buffaloes, sheep, birds, fish, and other plants and animals had become involved in multiple competing visions of wetland biopolitics. I spent my days with my hosts cooking, attending gatherings, playing with children, tending to water buffaloes, growing vegetables, and fishing in the delta's lakes. It was apparent that everyday rural livelihoods, including those of nonhuman animals, were affected by, and, in turn made claims on, wetland science and governance.

While in the Gediz Delta, I shadowed NGO and university scientists and lived in a seaside neighborhood in the nearby metropolis of Izmir. While in the Kızılırmak Delta, I lived and worked with rice cultivators, water buffalo herders, cash-crop farmers, and fishermen, all of whom were ambivalent about the new developments in the delta. Delta municipalities invested in ecotourism infrastructure and tore down "illegal" beachside vacation

Map 1. Turkey and neighboring countries, showing the main places mentioned in the text. Drawing by Benjamin Siegel.

houses; scientists obtained grants for their wetland research and for setting up wetland education activities. However, delta farmers worried about new conservation restrictions on their ability to graze their livestock; to collect reeds, wood, and leeches in the protected area; to fish; and to gain access to the beach.

The Infrastructure of Wetlands

"Wetland" is a very broad scientific and legal category that emerged out of concerns, coalescing in the mid-twentieth century, around the destructive environmental effects of resource extraction and conversion to farmland—a conversation that involved European and North American scientists, hunters, and birders (and largely excluded wetland-dwelling communities worldwide). The category "wetland" has been a semantic sponge, absorbing changing environmental preoccupations throughout the twentieth and twenty-first centuries: from waterbird habitat to biodiversity, water security, international development, participatory conservation, ecosystem services, and climate change.

Wetlands remain critical environments to think and to live with in the early twenty-first century. Places like the Gediz and Kızılırmak Deltas, and the many other wetland ecologies that have been drained, filled, and otherwise transformed to their disappearance in Turkey and elsewhere, make apparent the uncertain and precarious futures of human and nonhuman livelihoods. Where an ethnography of wetlands reveals the contradictory politics of environmental conservation, it also demonstrates the importance of attending to historically layered transformations of place and environments. The Gediz and Kızılırmak coastal delta wetlands are watery, terrestrial, and amphibious[7]—their materiality is not static but ever changing and in flow and very much at stake in debates among communities of farmers, fishers, scientists, bureaucrats, and NGO workers. In both places, the wetlands are also at once urban, industrial, rural, wild, and engineered.

Wetland environments are inseparable from the work of infrastructure. You find wetlands at the edge of agricultural fields, industrial ports, sewer overflows, abandoned industrial sites, city parks, and many other places. For environmental scientists, planners, and conservationists, the wetland has itself in the past two decades also become natural infrastructure for supporting the livability of Planet Earth. This book seeks to apprehend environmental infrastructure in its most expansive meaning. Infrastructures are material assemblages of things and people that move through space while remaking it, and also abstract kinds of calculative reason. Anthropologists studying infrastructures have often focused on the things infrastructures do, the social relations and moral subjectivities they produce, and their politics—whether planned or accidental. This approach derives from notions of infrastructure in earlier science and technology studies (STS) as invisible systems of scientific organization embedded in specific social arrangements and knowledge.[8]

From its late nineteenth-century use in French and English to indicate the material substrate below railway tracks, in the post–World War II era, infrastructure denoted the fixed installations of military deployments, such as airfields, signal communications, and headquarters. With the rise of "development" as a political model and an international form of intervention, the construction of infrastructure, large and small, became central to processes of modernization.[9] Infrastructure was simultaneously recast as the material precondition for, and a symbol of, industrialization, economic growth, and political power.[10]

For my Turkish interlocutors in the wetlands, infrastructures are material constructions, maintained through work, that always produce ecological effects and carry political meaning. During a conversation I had with a rice farmer about a small irrigation canal in the Kızılırmak Delta, the farmer connected the canal to the large-scale dams upstream on the Kızılırmak River; the relentless work of coordinating the flow of water across multiple fields; the high cost of the electricity pump; the flow of drainage waters to the delta's marshes; and the slow death of the delta's frogs, fish, and birds from the excess of fertilizers and pesticides that farmers are compelled to use to grow their hybrid rice varieties for the national market. Another friend noted that the number of birds had visibly increased two years after a wetland road was closed to cars. And fishermen in the Gediz Delta talked to me about the relentless traditional work of maintaining a "natural" lagoon, which involved moving stones and reeds to create favorable spawning habitat for their catch. Infrastructure is a useful concept for anthropologists because it is always relational. Infrastructure becomes visible when it breaks down, or in its absence. But even the most invisible infrastructure is visible to those who work to build it and make it work.[11]

Anthropological and historical accounts of infrastructure have been particularly useful in illuminating processes of political rule, community belonging, and resistance. The movements that infrastructures allow—of energy, capital, media people, and goods—also form the connective tissue of states and symbolize their power.[12] Failing urban water infrastructures reveal to their users centralized political systems that are unable to reach citizens, left to fend for themselves against breakdowns,[13] or cultivating new practices of care to tame unruly sewage and maintain everyday neighborly relations at the thresholds of their homes.[14] While access to water infrastructure may be central to informal settlers' claims of belonging to a city polity, it can also produce new moral subjects.[15] Infrastructures at once materialize processes of belonging, reproduce sectarian communities, and shape subjects as they move through them.[16] The materiality of infrastructure can work to support colonial projects of conquest and displacement.[17] Contestations over infrastructures, conversely, constitute new political connections and form novel collectivities; the communicative and collaborative networks people create and maintain, or work to disrupt and exclude others, become invisible "social" infrastructures.[18]

If infrastructure is an assemblage to make certain things move, one can ask what is moving, and what are the materials, technologies, peoples, ontologies, and knowledge that constitute the network? The answer is always shifting, as it depends on a situated perspective. Anthropologists have noticed that things—landfills, for example, discarded bread, or forests—can *become* infrastructural, as they conjure and facilitate practices and affective responses that generate their own patterns and social order.[19] This book builds on these analyses of environmental relations and transformations to call attention to the mutual constitution of ecologies and infrastructures as conduits for moral claims about more-than-human livelihood in uncertain times. Environments are constituted through layered histories of work and human-built infrastructures.[20] The concept of *environmental infrastructure* calls attention to the expert notion, arising in the 1990s from the work of ecosystem economists,[21] that ecologies themselves perform the work of human-built systems—a kind of work that can become commensurable in monetary terms.[22] Used as a tool for anthropological analysis, the notion of environmental infrastructure can also point to the varied ways in which "built" and "natural" environments are co-constituted and entangled.[23]

Wetlands illuminate these multifaceted aspects of infrastructure in, of, and as ecology. During my research, I learned to see the Gediz and the Kızılırmak Delta wetlands at once as sites of capitalistic speculation, objects of ecological care, open-air scientific laboratories, and experimental grounds for agro-economic development. They are shaped by the work and visions of university experts, municipal institutions, national ministerial offices, and transnational conservation protocol and inspire contrasting moral claims about translocality, expertise, and temporality. In contemporary Turkey, as in many other places, wetlands are an important site of everyday contestations—for middle- and working-class residents, scientists, bureaucrats, and farmers—over new and foreclosed possibilities for human and nonhuman livelihoods in a time of uncertain politics and in precarious and rapidly changing environments. I emphasize the *mutual constitution* of ecologies and infrastructures rather than their opposition, both as lived environments and as conduits for moral claims about valuations of human and nonhuman livelihoods in their ecological entanglements. These moral claims are often also violent and, to others, immoral. Moral ecologies reveal the work of power and inequality at play in everyday environmental politics, ethics, and practices.

Wetlands as Moral Ecologies

My biologist friends Emre and Deniz worked with great commitment and care to create more just wetland *doğa kültürü*, "nature-culture," as Deniz called it, or *biokültür*, "bioculture," as Emre and other friends of his would say. Their work is that of crafting a *moral ecology*. Moral ecologies are forms of ecological practice and thought in which morality—the concern with what is of value in life[24]—is at stake. Aspirations of justice and ethical subjectivity are staked on relations between people, plants, animals, fungi, water, and other organisms.[25] Moral ecologies are both *assessments* of justice and motivations for action. They help account for people's ethical impulses of caring for particular ecological arrangements: care that often results in violent outcomes—depending on the perspective. This analytic is helpful for understanding how, and why, people confront and respond to environmental transformations in various ways. I build on recent anthropological scholarship highlighting the interconnections of ecology and infrastructure,[26] as well as on notions of moral economy.[27] Moral ecologies, beyond their more common use to describe peasant, indigenous, and activist resistance,[28] or deployed interchangeably with the concept of moral economy,[29] are particularly useful for the anthropological study of environmental expertise, highlighting practitioners' ethical and affective commitments to their work. Moral ecologies are crisscrossed with and productive of politics and reveal the complex ways in which practices of ecological care, conservation, and love can at the same time also involve violence, dispossession, and marginalization of unwanted people, organisms, and ecological relations.

International journalistic reporting has largely portrayed environmental mobilizations in contemporary Turkey as a stark leftist and secular opposition to its current ruling party, the Muslim conservative Adalet ve Kalkınma Partisi (AKP). Indeed, many Turkish environmental activists vocally opposed to quarries, mining, nuclear energy, hydropower, and deforestation have faced an intensification of violent political repression over the last decade. However, the multiple and contrasting moral ecologies at play in Turkish wetlands are not a simple microcosm of the divisive politics of the "New Turkey," a term first used by AKP leaders and later appropriated by its opponents as a term of critique for the regime.[30] In recent years, and particularly after the Arab Spring of 2011, political commentators have frequently debated whether Turkey offers a model of moderate Islamic democracy and balanced neoliberalism. Some journalists have positioned Turkey's political mobilizations as exceptional; and others, as a template for regional politics. This postulation is not new. Throughout the twentieth century, social scientists have studied Turkey as a material laboratory for theorizing and measuring modernization, whereby Turkey was posited as either exceptional or a model for other political systems.[31]

Wetland conservation practices in Turkey, however, are not a clear-cut critique of authoritarian rule, nor are wetlands symbolic sites for displaying governmental might and party politics. Rather, in their contestation over how to preserve and manage wetlands, and how to live in them, different social groups wrestle with what it means to be moral ecological subjects at a time of political and climatic uncertainty. Some of my urban friends in Izmir and Istanbul suggested that wetlands showcase the ongoing battle between an aggressive governmental agenda for rapacious development at all costs and demands for ecological justice espoused by ordinary people. I came to disagree with them. I do not take wetlands as simple stand-ins for the polarizing politics of Turkish nature. Instead, I situate wetlands as part of specific provincial and municipal administrative units, bordering villages, towns, and metropolises and shaped by centralized planning, international currents, and local politics.

Where my interlocutors have seen in the wetlands flourishing models for Turkish democracy, resistance, hope, and sustainable futures—and have worked to realize them—I have tried my best to follow them in their visions and partake in their everyday work so that I could better understand and write about it. Whereas the wetland is, for some of my interlocutors, a tangible symbol of the structural failures of Turkey's environmental policy, I have attended to this perspective while trying to convey the multiplicity of contrasting perspectives on what exactly constitutes failure. Everyday practices of work and politics in the wetland, however, are more directly connected to its ecological flourishing. I heard fishermen puzzle over a newly introduced species of lake fish that is eating their catch, learned to marvel with birders as flamingos performed their seasonal courtship dance, followed photographers looking for the spring blossoming of floating water lilies, harvested small handfuls of Salicornia and wild asparagus for the evening's dinner with field biologist friends, and rejoiced with farmers at the birth of a new gruffy water buffalo with an endearing personality.

In the past two decades, Turkey has become a productive field site for nuanced anthropological inquiries of the nation-state; religion and secularism; urbanization; kinship and gender; and ethnicity, war, and migration. The remaking of agro-economic landscapes and ecologies has been central to Turkish nation-state building and its discontents; however, ethnographers have largely sidelined Turkish environments as blank slates, existing only in the background of wider political and economic processes. Ottoman and Turkish historians have begun writing on climate histories, forestry politics, genocide environments, disease landscape, water, and animals.[32] Recent environmental ethnographies of Turkey have centered on environmental mobilizations, focusing on resistance to the construction of new energy infrastructure, such as nuclear, coal, and hydropower plants.[33]

Scholars have suggested that Turkish grassroots environmentalism attracts different constituencies around a common political "malcontent" for which the environment serves merely as proxy.[34] Alternatively, others have posited that, despite environmental advocates' proclamation of being "above politics," environmentalism only reproduces rather than challenges class and political divisions,[35] or that it results in a legitimization of state governance.[36] A critique of these approaches is that they render lived environments marginal to, and often overdetermined by, other politics.[37]

Sometimes I observed in my research that wetlands are indeed proxies reflecting other political concerns, and class and political positioning may in fact be reproduced in conflicts over wetland management. However, portraying wetlands as just reflective of other political formations would result in perpetuating a structural marginalization of the environment as a secondary concern, a strategy that is also often used by political elites themselves to suppress people's claims for environmental justice and moral ecologies. In contrast, this book analyzes how wetlands are made—and the varied crafting of moral ecologies.

Conflicted Matters

Over the course of my field research in the Gediz and Kızılırmak Delta wetlands, between 2012 and 2018, I would learn that people who lived and worked on wetlands—particularly scientists, farmers, fishers, and bureaucrats—understood them as complex lived ecologies shaped by cultural and political forces. They were environments requiring constant work of care, variously directed at maintaining wetland ecologies; cultivating them; rewilding; rendering them productive; studying, transforming, or governing them.

Where water should flow, for example, what kind of water, how much, what kind of flow, and what new configurations of more-than-human livelihoods would result from these decisions were very much contested matters.

On that hot summer day in 2012, in the abandoned saltpan, Emre was working on a vegetation map, tracing the expansion of reeds, which he used as an indicator for the moving boundary between fresh and saltier water. The "reed line" had been gradually advancing toward the Aegean Sea, taking over the saltmarsh habitats that thrived in the lower delta. This was an effect of recent wetland management: to counter decades of drought and increased water extraction for agriculture, the wetland management agency, where Emre worked, regularly purchased water from the delta's irrigation cooperative to, literally, water the wetland to prevent the marshes from drying. However, irrigation also transformed them into freshwater environments, and more toxic agricultural and industrial runoff seeped into the saltmarsh ecology. Wetland materialities like these were always contested and generative of moral assessments of ecology. A senior university biologist supported irrigating the wetlands as a palliative for the increasing violence of summer drought and as a way to counter the effects of agricultural water use. But Emre and others envisioned this intervention as reducing the saline marsh habitat. They fantasized about restoring the ebbs and flows of the lower delta before dikes and irrigation canals had turned it into drier, more static earth, starkly separated from the marshes and the sea.

I encountered many contrasting and conflicting visions of the wetland in both the Gediz and Kızılırmak Deltas. Each wetland imagination also entailed a material transformation of the wetland to produce or support specific ecological relationships, all of which also included people. University scientists envisioned turning wetlands into university laboratories—a transformation that involved denying access to local fishers, herders, and foragers—and also controlling the population of eucalyptus trees, feral horses, buffaloes, invasive fish, and other problematic species. City officials saw the wetland as a potential tourist attraction. Urban planners worked on projects that highlighted the functions of the wetland as a natural and "green" infrastructure, one that would keep Turkish metropolises, towns, and villages livable for future generations. Real estate speculators saw the wetlands as a site to be drained for prime real estate as the city expanded northward. For international scientific teams, the deltas were also a node in the international network of wetland managers and conservationists. In each of these visions, notions of ecological relations were connected in various ways to moral aspirations for human live-lihood and politics. Rather than serve as a simple reflection of the social order, the wetland itself muddles, sediments, and transforms the political.

The Invention of Turkish Wetlands

As Turkish bureaucrats and officials appropriated the category "wetland" in the second half of the twentieth century, they embarked on a project to name, catalogue, demarcate, evaluate, study, manage, and preserve these ambiguous areas saturated with water. On February 2, 1998, Turkey's president, Süleyman Demirel, gave a speech at a wetland conservation conference organized by the Ministry of Environment and the Society for the Conservation of Nature (DHKD) on occasion of World Wetlands Day. Demirel had trained as a civil engineer, had worked on dam and drainage projects, and then was appointed head of the State Hydraulic Works (Devlet Su Işleri [DSI]) in the 1950s and 1960s. In his speech, he reflected on this work: "In those days, our priority was to eliminate malaria. So, what did we do? We drained every swamp [bataklık] we could find. Of course, while doing that, we also damaged nature." But now protecting Turkey's "beautiful wetlands" (sulakalan) and nature (doğa) had become a national duty, and "we will protect [nature and wetlands] in the same way we protect our borders, everything in our country, its value, our honor and dignity."[38]

Demirel was Turkey's prime minister in 1994, and in this role he presided over Turkey's joining the Ramsar Convention for the Conservation of Wet-lands of International Importance. Turkish politicians, bureaucrats, and sci-entists had been involved with international conversations over wetland con-servation since the 1960s. Demirel's speech demonstrated that the wetland was no longer just a malaria-ridden environment of unruly populations. It had become a subject of state governance and also a material proxy for nation-alist concerns over the protection of borders, women, national honor, and racial and ethnic purity. However, "swamps" to be drained were not simply replaced by "wetlands" to be preserved.

The trope of swamp drainage giving way to wetland conservation also reveals the ways in which Turkish swamps and wetlands have been copro-duced. Wetland conservation continues to be actualized in ways that are con-nected to, and do not simply oppose, the drainage, reclamation, flooding, and control of swamps and marshes. Throughout this book, I suggest that

contemporary wetland conservation in Turkey and elsewhere has often rein-
vented and repurposed older tools of swamp reclamation to legitimize technocratic environmental management that romanticize human and multispecies wetland livelihoods, while also precluding their possibility. At the same
time, residents, scientists, and bureaucrats have also found in the wetland a
fertile yet uncertain, murky, and mobile ground for cultivating new aspirations of democracy, multispecies livelihoods, and more just ecologies.

Starting in the 1960s, the question of wetland conservation was framed by
wider debates over Turkey's development. In 1967, Ahmet Varışçığıl, representing Turkey's DSI, spoke at a wetland conservation conference in Ankara.
He underlined the economic benefits of the drained wetlands, which he estimated at 150,000 hectares: they had been developed for agriculture and rice
cultivation, planted with eucalyptus trees, and used for fishing, pastures, and
game reserves. Rapid population growth and the need for new settlements,
he averred, had forced Turkey to abandon older and traditional wetland uses,
such as reed cutting, fishing, and grazing.[39]

At the same conference, Swiss scientist and conservationist Luc Hoffmann
warned against the drainage of wetlands for agricultural development, which
could result in long-term loss of productivity. He urged leaders in Turkey and
other Middle Eastern countries to instead plan for the "improved utilization"
of wetlands, integrated in development planning: utilizing them for flood and
water table control, for instance; and for timber, fish culture, reed production, fur hunting, wildfowl; and for scientific, educational, and recreational
purposes.[40]

Three decades later, the authors of a book on Turkey's Important Bird
Areas (IBAs) diagnosed that "the loss and degradation of wetland and other
habitats continues at an alarming rate . . . compounded by the lack of political will to tackle environmental issues and the complicated and contradictory nature of existing legislation."[41] They calculated that at least 1.3 million
hectares of wetland habitats had been lost as a result of flood protection, reclamation, damming, water diversion, and other water-manipulation schemes
in the twentieth century. Even more wetlands had disappeared after the diversion or damming of rivers and streams or were flooded in the basins of dam
reservoirs.[42] In this, Turkey was not exceptional. By the early 1990s, a field of
"wetland loss studies" had emerged to calculate and document the worldwide
destruction of wetlands and, in a conference proceedings, a British wetland
scientist reported that "the most common type of Mediterranean wetland

is the lost wetland."[43] At the turn of the twenty-first century, UNESCO announced that about 50 percent of the world's wetlands had been lost, mostly converted to agriculture.[44]

In 2012, Turkish media outlets, drawing from a Turkish environmental NGO's press conference, reported that Turkey had drained about two million hectares of wetlands since the 1960s.[45] This news has been widely broadcast each subsequent February 2, on the occasion of World Wetlands Day, by media channels across the political spectrum. But Turkish government officials have also contested this estimate. In 2017, a regional director of the Water and Forestry Ministry (renamed in 2018 the Ministry of Agriculture and Forestry) claimed that "there has been *no change* in Turkey's wetlands since the 1950s" and that changes in water levels had to do with seasonal fluctuations and climatic changes. He also leveraged the very broad definition of wetland used by the Ramsar Convention to challenge the environmentalists' concern with the damaging effects of water dams. "Dams are important habitats for wildlife," he said. "And the Ramsar definition includes 'man-made wetlands,'" like the contested dam reservoirs.[46] The concern with wetland conservation in Turkey, as in many other places in the world, has become one bringing together scientists, environmental advocates, international experts, and national bureaucrats over shared questions of wetland conservation, with many disagreements over what should be preserved, how, why, and for whom.

Across Two Wetlands

How do people and other nonhuman beings inhabit, make sense of, and contest wetland environments? How is the wetland denomination shaped by longer regional environmental histories? How was the "wetland" category invented, and how did it become such a powerful global object, strategically interpreted and applied to different contexts? I have pursued these three questions iteratively because I see wetlands operating at once as sites for moral claims about multispecies livelihoods and ecology, scientific categories and laboratories, and mobile environments formed through histories of resource use, infrastructures, and environmental imaginaries. Throughout my research, these three questions kept building on each other, shaping how I moved through different sites, communities, methods, and materials, and the decisions I took.

My position as a young southern European woman pursuing her doctorate studies in the United States helped open many doors and facilitated new
connections and friendships, but it also kept others shut. I could not always
get the access I hoped to many a wetland expert, an environmental advocate,
a community of wetland residents, or an archive. And there were countless
occasions when I retrospectively wished I had asked different questions, pursued conversations in new directions, or followed a lead to a place, a person,
or a paper trail of documents that at the time I thought was too far-flung to be
related to my research.

In August 2012, I had recently arrived in the city of Izmir. Two weeks earlier, at an environmental NGO's headquarters in Ankara, a feminist environmental advocate had suggested that to understand the politics of water in Turkey and the ways water becomes a contested resource, I had to follow the flow
of water itself, from mountain springs to the plains and then to the coastal
deltas. She had been working on wetland programming for the NGO. Wetlands, she told me, are where it all comes together. In the wetlands, I learned,
one can understand the relationships between conservationists, industry,
agriculture, the layers of water governance, and entangled topology she called,
in Turkish, "nature-culture."

The NGO scientist suggested I look at the fourteen Turkish wetlands on
the Ramsar list of Wetlands of International Importance and that I focus on
places on which environmental NGOs and governmental agencies had worked
extensively to implement projects of wetland governance and conservation.
She invited me to follow her and her team on a short trip to Burdur Lake,
one of the Ramsar wetland sites, and she put me in touch with the NGO's
coordinator for the Gediz Delta in Izmir, Deniz. I also got the contact information for a well-known ornithologist in Samsun, who had been working on
ornithological research in the Kızılırmak Delta and had collaborated with the
NGO on wetland conservation efforts over the past decade. In the summer of
2012, she hosted me for three weeks to take part in the seasonal ornithological research and bird ringing and introduced me to her collaborators at the
university, her students, and rural delta residents.

I returned to the Turkish wetlands a year later. In 2013, during my first few
weeks shadowing Deniz and other wetland advocates in Izmir, I noticed that
they frequently talked about the Kızılırmak Delta. In a way, the deltas were
similar: large agrarian plains with villages, towns, and cities and near a large
metropolis (Samsun had just under a million residents, and Izmir over four

million). Both deltas had been cosmopolitan until the early twentieth century, when, between 1914 and 1923, Orthodox and Pontic Greek and Armenian communities were violently killed, transferred, or removed.[47] Their histories have largely been elided, and their houses, churches, and schools destroyed or repurposed. Both deltas became resettlement sites for countless waves of Muslim refugees and migrants during the late Ottoman and early Republican periods—and during the time of my fieldwork they were populated in large part by their descendants.

Their transformation of the deltas from marsh and swamp into wetlands was also similar. The Gediz and Kızılırmak Delta wetlands had been added to the Ramsar list of Wetlands of International Importance in the same year, in 1998. Many rural residents in the Kızılırmak Delta I talked to were aware of, and concerned about, the restrictions to livestock grazing in the Gediz Delta's conservation area. In 2016, bureaucrats working in the Samsun Province implemented a wetland management structure for the Kızılırmak Delta, adopting a structure copied from the one in place for the Gediz Delta. In turn, Gediz Delta advocates applied for UNESCO World Heritage status when the Kızılırmak Delta entered the UNESCO's provisional list in 2017. Many NGO experts working in one delta had been involved in projects of scientific research, advocacy, and wetland management in the other. Birds, and the people who study and care for them, played a central role. Turkish birders and ornithologists were very familiar with both deltas, as they were popular nodes in the transnational migratory routes of Turkish birds and habitats for hundreds more nonmigratory species. Bureaucrats also traveled between the two sites to attend wetland conferences and meetings.

I decided that I would explore these connection between the two deltas. I rented a shared apartment in an old house in Izmir's Alsancak neighborhood, formerly an Italian neighborhood called La Punta, just a few blocks away from the NGO's headquarters, where I spent three months as a volunteer. I traveled to the Gediz Delta in different ways. Often, I would catch the ferry across the bay and then take a bus or ride my bike. Other times I would get a ride with the management center's staff from Izmir. A municipal bus connected the suburban train line to the delta's villages, and many friends hosted me in the delta.

In Izmir, I spent weeks at the National Library, in the busy market district of Kemeraltı, to research the history of the Gediz Delta. Friends introduced me to an urban and environmental sociologist at a local university who hosted me in his department for an academic year and introduced me to the many natural scientists who were working on varied research projects in the Gediz Delta. I also observed the work and conversations of a team of scientists from the French research institution Tour du Valat during two of their periodic visits to the Gediz Delta wetlands, where they worked as part of their Mediterranean wetland management and research efforts.

At the National Wetland Conference, hosted in Samsun in October 2013, I met communities of scientists, state bureaucrats, and environmentalists advocating for the conservation of the Kızılırmak Delta. The university ornithologists invited me to join the seasonal bird-ringing camp in the delta for three weeks. A university agricultural engineer helped me find a family of delta farmers who would host me in their village the following summer. In Doğanca, a municipality of Bafra, four different families hosted me between 2014 and 2018, for periods ranging from a few days to several months, and allowed me to work alongside them, collect oral histories, and partake in family and religious celebrations.

While living in the Kızılırmak Delta, I was able to join several wetland education summer camps for university students and field trips for scientists and ordinary citizens, and I participated in wetland management meetings with scientists and bureaucrats as well as a civil-society group. I also helped university scientists envision a year-long participatory conservation approach to the delta's wetlands and obtain a major European Union (EU) grant to support it, though the project collapsed because of a political clash between the university and one of the collaborating NGOs in the aftermath of the July 2016 coup attempt.

During this time, I also traveled to Switzerland to visit the Ramsar, International Union for Conservation of Nature (IUCN), and WWF offices and their archives for two weeks. I consulted the Asia Minor archive in Athens to find memories of the displaced Pontic Greeks. And I spent a month in Ankara working in the Republican Archive on wetland drainage histories, interviewing bureaucrats at the Wetlands Bureau, part of what was then the Ministry of Water and Forestry, and gathering wetland conservation and management documents produced by the Turkish State. In the two deltas, the visions for wetland conservation, and the politics it engenders, have continued to change since I left the field. Despite these changes, wetlands continue to be sites of hope, struggle, and contested moral ecologies in times of political and environmental uncertainty.

Connections: Fieldwork and Paper Trails

To understand the ways wetland designations helped transform material environments and their political ecologies, I followed the networks through which wetland advocates and experts understood their work: patchy assemblages of documents, ideas, people, nonhuman species, manuals, regulations, institutions, funds, and projects.[48] These were ones that connected Turkey's distinct wetland sites to one another, through practices of comparison, similarity, and abstraction.[49] These practices knitted distinct Turkish wetlands together and enabled certain kinds of moral and political action on behalf of wetland ecologies. During my research in wetlands, cities, and archives, I realized that the social, political, scientific, and ecological relations that emerged between and connected different wetland sites answered some of my early research questions—how the wetland had become a meaningful category for Turkish environmentalists, bureaucrats, scientists, and residents.

The wetland, I learned, is always connective. The wetland categories that became hegemonic in the late twentieth and early twenty-first centuries are all-encompassing and invite connections and comparisons across far-flung places.[50] As I detail in Chapter 1, wetlands were invented as a tool to coordinate the preservation of waterbird stocks across state and national borders in the early decades of the twentieth century. After the 1960s, wetlands remained connective categories, gelled through varied international conferences, conventions, treaties, and programs. No wetland exists alone.

Studying the wetland anthropologically required crafting a toolkit of different methods, theories, and literatures. It also meant finding ways to creatively account for wetlands' materiality, ecology, and temporality. I moved somewhat fluidly through different locales and communities while also forging lasting collaborations, friendships, and relations of mutual trust with NGO workers, Turkish and foreign scientists, farmers, urban residents, and bureaucrats. I recorded a few dozen interviews and transcribed many more informal conversations. I tended to water buffaloes, learned to recognize wetland birds, and acquired some basic fishing skills.

From the very first few weeks of my research, my ethnographic note taking described activities and everyday practices in the wetlands and conversations with my interlocutors over a shared meal and shared work, which I came to prefer over more formal interviews. This choice is reflected in the ethnographic material I draw on in this book. This is how I learned that farmers, scientists, NGO staff, and bureaucrats' invocation of the wetlands was used for strategic and contingent goals and that "wetland" was taken to mean wildly different things. The notion that claims about wetlands are often ones of *moral ecology* emerged from these ethnographic encounters and collaborations.

To write about the history of wetland denominations in the context of environmental and agricultural transformation, I read documents on dredging and agricultural development in the Republican Archives in Ankara, explored the Asia Minor archives in Athens for firsthand accounts of displacement, recorded oral histories in my field sites, and read the work of Ottoman and Turkish environmental historians. Environmental historians have taught us that, all over the world, marshes, swamps, lagoons, bogs, and other places of shallow, saturated water in seasonal flow are often hard-to-conquer places where populations found refuge to protect themselves from the reach of empires and nation-states. State-sponsored draining of swamps and wetlands was more than an economic project to produce more arable land. It was also concurrent with attempts to create legible, sedentary, taxable, and acquiescent populations.[51]

Nomadism and Wetlands

The draining of Ottoman swamps and marshes, an effort that intensified in the nineteenth century, was aimed at resettling stable, and more easily taxable, agricultural subjects in the plains formerly inhabited by seasonal groups of hunters, fishers, and pastoralists.[52] In the early years of the Turkish republic, marsh drainage intersected with projects of public health that sought to eradicate malaria to create new healthy citizen bodies and productive landscapes. Building the small town of Ankara into a modern planned capital for Turkey, for example, required draining the marshes around the city.[53] But in countless other places in the country, lakes and delta marshes were drained to create new agricultural land and settlements. Drainage became an established mode of engineering economic growth and modernist development.

In the 1960s, Turkish scientists and planners began to embrace new visions of wetland conservation—the wetland had become an open-air laboratory for ornithological and ecological research. In his historical account of wetland conservation in the United States, Robert Wilson found that wetland refuges were created alongside and used the same infrastructure that supported irrigated agriculture.[54] Rather than understand wetland conservation as a palliative of and opposed to drainage, in this book I understand them as similar and interrelated projects of state political control.

good use of comparison

In STS scholarship, a "boundary object" is a scientific thing around which groups of experts with opposing perspectives are nevertheless able

Science, Technology and Society

to collaborate, even across diverging interpretations.[55] How did the wetland become such a vital boundary object and a powerful global object, strategically applied by different communities of people to such different contexts? This question led me to trace the debates and scientific work that led to the invention and stabilization of the category "wetland" in the twentieth century. There was no ready-made "wetlands of Turkey archive" I could consult, so I followed connections, intuitions, and leads to different places to create my own. Initially, I expected this material would just supplement my ethnography. As I kept building my collection, I began to see a meandering story worth telling through my reading of these documents.

The two weeks I spent exploring libraries and archives at the IUCN, Ramsar Bureau, and WWF in Gland, Switzerland, in April 2014 revealed a wealth of institutional documents and reports that helped me trace the birth of the wetland in a series of scientific meetings and conferences, including the Ramsar Conference of 1971, the first international agreement on the conservation of a specific ecosystem. I also benefited from field reports and other documents that staff at those institutions made available to me, including their personal accounts of memorable turning points of wetland conservation. The Ramsar office's "Turkey" box, while disappointingly small, contained much of the material that Turkish ministries had published in the 1990s and early 2000s and mailed to Switzerland, including site-specific pamphlets for each of the country's Ramsar wetlands, as well as descriptions and lists of the wetlands of Turkey and of the country's conservation efforts. The French wetland research institute Tour du Valat offered a good online archive of its more recent research undertakings, and the Tour du Valat team in Izmir shared with me their project reports and scientific findings. To trace the early twentieth-century history of wetland conservation, I scoured university libraries in both Turkey and the United States to obtain publications and reports from European waterfowl organizations, ornithological publications, scientific wetland conference proceedings, and the wetland reports of the US Fish and Wildlife Service.

I followed a similarly convoluted trail of papers on Turkish governmental and NGO wetland advocacy work, which only began to make sense to me, as intersecting narrative threads, after I finished my fieldwork. In Ankara, staff at the wetland office, which at the time of my visit was located within the Ministry of Water and Forestry, shared with me their published material since the 1990s. University scientists in Izmir and in Samsun gave me copies of their own material, including their scientific articles and work in progress, PhD theses, management plans, maps, newsletters, and legal documents. Participating in the third National Wetland Conference, hosted in Samsun in 2013, helped me gather newer material, and I have kept updating my working archive until drafting this book. Friends I made during my research at various Turkish environmental NGOs allowed me to consult, borrow, and copy their scientific reports, publications, lawsuit documents, photographs, and maps.

Chapter Overview

Wetland conservation today presents itself as global in scope, addressing threats to biodiversity, climate, and livelihood at the planetary scale. Chapter 1 analyzes the emergence of the category "wetland"—as both an ecological term and as a subject of science and conservation—in the work of Turkish, European, and North American scientists and environmental advocates throughout the twentieth and twenty-first centuries. In the early twentieth century, North American and European scientists debated legislation to protect cross-border bird migration. Since then, the new ecological and legal category "wetland" has come to encompass bogs, swamps, salt marshes, mudlands, and other areas of saturated land. Through intergovernmental agreements such as the 1971 Ramsar Convention for the conservation of wetlands, environmental organizations have sought to convince governments to stop the reclamation of wetlands for agricultural, residential, and industrial use. The ways in which different Turkish actors have made wetlands have been neither a wholehearted endorsement of a national, central state nor necessarily an ideological rejection. Instead, Turkish wetland conservation is a pragmatic amalgam of strategies in pursuit of contingent goals.

The following chapters focus on the varied ways wetland abstractions have been transformed in contestations among different social groups over the moral ecologies of wetland livelihood in the Gediz and Kızılırmak Deltas. Chapters 2 and 3 analyze contested moral ecologies of water, sediments, and infrastructure in the Gediz Delta. Chapters 4 and 5 attend to scientists' care for the Kızılırmak Delta and to the effects of wetland regulations on the livelihood of delta residents—farmers and fishers, and also including birds, fish, and bovines.

Drawing from ethnographic research among environmental NGOs, residents, scientists, and state officials in Izmir, Chapter 2 analyzes the contested nature of the Gediz Delta wetlands—variably envisioned as ecological and cultural spaces of human and nonhuman livelihood in flow. Following Turkish scientists at work through muddy estuaries, in field stations, at university international conferences, and in their homes, this chapter frames the competing ideas of wetlands produced through different encounters. Like the materiality of wetland waters, flows, and sediments, their moral ecologies—human and nonhuman alike—are always contested and in the making. For example, in 2014, wetland managers in the Gediz Delta paid an irrigation cooperative to, literally, water the wetland, debating among themselves and others the desirable qualities, quantities, and flows of water. In deciding how much irrigation water to purchase to sustain particular wetland ecologies, wetland managers advanced contrasting understandings of the animal or plant species that would thrive or die. Likewise, residents reckoned with the new kinds of habitats produced by the infrastructures of factories, high-rises, wastewater plants, and dredging boats. As a wetland official secretly helped a village head petition for grazing rights in the wetland area, and contravened his mandate to fine herders, he both performed and circumvented bureaucratic structures of wetland conservation. This chapter reveals the sticky sediments and moral quandaries of wetland science in the making.

Zeroing in on an old fishing lagoon and on a new large-scale bridge project over the Bay of Izmir, Chapter 3 analyzes contemporary sites of conflict over the moral ecologies of wetland infrastructures and materialities. In 2014, a fishers' cooperative adopted the language of international wetland conservation to make demands about a decades-old conflict over fishing rights. As fishermen revisit old struggles in new deltaic ecologies, environmental activists resist the encroachment of planned megaprojects, and university researchers draw up plans for the "wetlands of the future," they advance vernacular notions of infrastructure sutured to specific ecologies and livelihoods. In 2013, residents of Izmir organized public demonstrations in defense of their living space and that of the city's birds—only to be silenced again after a 2016 coup attempt. These overlapping moral ecologies emerge from specific wetland encounters and underwrite authoritarian and emancipatory political projects alike. The cultural and political salience of wetland spaces in Izmir is enacted through responses to rapid transformations of the built environment in urban, industrial, and rural locales. I foreground the varied ways in which, for residents of the Gediz Delta, infrastructures and ecologies are mutually produced.

Since the 1990s, farmers, scientists, and bureaucrats have come to reimagine the Kızılırmak Delta, on Turkey's Black Sea coast, as a valuable ecology. Chapter 4 analyzes the ways in which these ecological imaginations of the wetland both include and exclude rural residents. A debate between fishermen and rice growers about fires in the lakes' reed beds, for instance, was not only about specific forms of water ecologies but also about who should make this decision and the effects on others' livelihoods. Through their everyday work practices, scientists come to care affectively for the delta as a wetland—and, in the process, become advocates for alternative futures and moral ecologies. The chapter also details the different ways in which farmers tend to their agrarian delta. These contrasting moral ecologies in a shifting environment construe new notions of livelihood. In engaging critically with wetland conservation, urban and rural residents, scientists, and bureaucrats alike also articulate new imaginations of the Turkish nation-state and moral notions of governance.

Projects of wetland conservation—stewarded by the state, universities, and NGOs—make claims on environments that have themselves been constituted by longer entanglements of crops, markets, infrastructure, and nation making. Chapter 5 examines how these transformations become entwined with the lives of nonhuman animals in the Kızılırmak Delta. As they decide what constitutes "wildlife," wetland managers, state officials, and scientists locate animal agency in the relationships they posit between animal bodies and ecologies. In contrast, farmers understand domestic and wild animals as intimate companions central to livelihood: active participants in the making and remaking of land and water. Rapid environmental and agro-economic changes in the delta in the late twentieth and early twenty-first centuries—expanding rice fields, disappearing forests, new drainage infrastructure, chemicals, and subsidies—have rendered certain animal populations "out of place." This shifting enactment of place affects animals in different ways; accordingly, a broader understanding of environmental change and sovereignty has become displaced on discussions of those species that belong in the delta within specific ecological and economic relations.

The Conclusion returns to wetlands' moral ecologies and questions about materiality in light of ongoing political uncertainty, authoritarianism, and displacement. In an era of rapid climate change, wetlands have become increasingly important sites for crafting environmental futures. I discuss the changing salience of wetland livelihoods in today's Turkey, particularly after the 2018 advent of a new presidential system paving the way for increasing authoritarianism and concentration of political power. I also reflect on my research interlocutors' theorizing of new possibilities and futures for Turkey's moral ecologies in and beyond the wetlands.

1   The Wetlands of Turkey

Convergence

On October 9, 1967, Turkey hosted, for the first time, an international wetland conference. Kemal Kurdaş, a former minister of finance and International Monetary Fund (IMF) economist, who now served as president of the Middle East Technical University in Ankara and vice president of the Turkish Association for the Conservation of Natural Resources, gave a keynote speech. The Anatolian soil had been subjected to reckless agricultural exploitation for millennia, Kurdaş declared, appealing to the Turks' civic duty to protect their natural heritage. The conference, he emphasized, was evidence of the international cooperation supporting Turkish efforts of natural resource conservation. In another presentation, Ahmet Varışlığıl of the DSI,[1] described the pivotal role played by the Turkish government in draining wetlands to create new profitable agricultural land to redistribute to farmers.[2] Turkey had 350,000 hectares of very "typical" and "widespread" wetlands, he reported. Half of these had already been drained, and the remaining had been surveyed and were scheduled for drainage.[3]

The conference made apparent a novel convergence between two trajectories. First, Turkish officials, like those of most other states and imperial powers, had long recast marshes and swamps as unhealthy environments to be reclaimed for agricultural and urban development. This was a late-Ottoman vision of sedentary agricultural civilization that had been implemented as national development during the first decades of the Turkish republic.

Second, and in stark contrast, since the early twentieth century, in response to large-scale drainage, birders and scientists had advanced an ethos of wetland conservation. This effort had produced new legal structures and scientific concepts of the wetlands in a mushrooming network of conservation and research institutions in Europe and North America. In the years and decades following the 1967 conference, Turkish birders—amateur and professional, working for the government, in universities, and in civil-society organizations—would become stewards of wetland protection in the country as the concept of the wetland came to encompass environmental conservation beyond the preservation of waterbird habitats.

At the 1967 meeting, the DSI bureaucrat cast Turkish wetlands as national and natural heritage to be conserved, but the economist envisioned them as the raw material of national development. Both men understood wetlands as national resources and even used the same Turkish neologism to describe them: *sulakalan*, a term that encompassed different ecologies saturated in water, from lakes to marshes, lagoons, volcanic pools, bays, and even dam reservoirs. The relentless work of simultaneous translation across English and Turkish that Resan Taşçıoğlu, an Ankara-based social worker and one of only two women at the conference, undertook throughout the conference discussions established the Turkish neologism for "wetland." *Sulakalan* meant, literally, "watery place." *Sulakalan* overlapped with but never fully replaced the older term *bataklık* (or *batak*), "the sinking place." *Bataklık* referred to swamps and marshes, carrying a negative connotation of them as unproductive and unhealthy places to be drained and reclaimed into agricultural land or urban development.[4]

In the 1960s, Turkish bureaucrats and scientists invented Turkish "wetlands."[5] Material transformations of Turkish watery environments converged with the legal and scientific development of concepts of the wetland, which Turkish state officials, birders, and scientists variously embraced and made their own. This convergence resulted in new categories of wetland governance and resources, which never fully supplanted but instead existed alongside visions of marsh and swamp reclamation. It also generated new environmental forms, whereby wetland preserves would coexist with large-scale agricultural reclamation projects, even sustained by the same hydraulic infrastructure.

Interest in wetlands as special kinds of ecologies began at the turn of the twentieth century with North American and European hunters' and scientists' concerns with the well-being of birds breeding and feeding in areas of

shallow water. In Turkey, European, North American, and Turkish conservation, scientists created and embraced categories of the wetland against the
backdrop of the large-scale drainage of places saturated in water. Conservationists, then as now, understood wetlands as ecologies that exist within international networks, since wetlands were connected by the migratory routes of
birds spanning national borders.

The wetland, as a concept, was stabilized and naturalized over the twentieth century through countless scientific publications, international conferences, collaborations, and treaties, as well as national legislation and nongovernmental activism. These wetland denominations have been semantic
sponges, absorbing changing notions of environment, politics, and conservation. I propose that wetland conservation did not pose a radical critique of
drainage but, rather, generated a complementary vision of state-led control
of unruly marshes to be transformed into scientific laboratories for wetland
research—sites for producing national and global nature.

The history of environmental conservation of wetland areas and other
nature preserves in the twentieth century is also part of a history of colonial
appropriation and dispossession of indigenous communities and politics.[6]
Nature conservationists turned lived landscapes into "wilderness" meant to
be enjoyed by elites as hunting and recreation grounds and as scientific field
sites. The exclusion of local populations from these restricted areas, often
motivated by other political concerns of territorial control, was openly justified by arguments about scientific research and species conservation.[7]

Taming Swamps: Drainage and Resettlement

The draining of marshes and swamps in the late Ottoman Empire and after
the founding of the Turkish republic was entangled with the varied aspirations of empire making, political control, and economic expansion. These
were realized through the reordering and control of "unruly" watery environments and their human, and nonhuman, populations. In the Turkish Republican era (after 1923), the control and drainage of marshes became central to
national development planning. Drained marshes were also the stage for the
biopolitical creation of new Turkish subjects in the aftermath of the killings
and forced deportation of Assyrians, Armenians, and Orthodox and Pontic
Greeks, alongside population exchanges that sought to produce ethnically,
racially, and religiously homogeneous Turkish citizens.[8]

In the nineteenth century, the Ottoman Empire was undergoing territorial loss and, simultaneously, an influx of new populations—predominantly displaced communities of Muslims from the Caucasus, eastern Europe, the Balkans, and the Mediterranean Islands. The resettling of migrants and refugees also coincided with imperial officials' attempts to sedentarize nomadic pastoralists and curtail their sovereignty. Ottoman statesmen envisioned sedentary cultivation as a harbinger of civilization and progress, while also eliminating malaria, a parasite carried by mosquitoes that thrived in the wet lowlands. An Ottoman doctor, Feyzullah Pasha, taught his students that "malaria likes unworked lands and desolate, empty countryside. It cannot hold up in the face of civilization and the efforts of mankind."[9]

While nearly all Ottoman wetlands were marked for drainage by the end of the Ottoman period, reclamation was difficult to implement: it was an expensive undertaking and proved ineffective in eliminating malarial ecologies.[10] The migrants and nomads who were relocated to permanent settlements on drained lowlands and lakes throughout the Ottoman Empire still frequently suffered from malaria epidemics. For example, almost all of the Circassian refugees who had been resettled in the Kızılırmak Delta after 1864 perished of malaria, after the government authorities declined their request for land in the Nebiyan Mountains. For this reason, communities often resisted: for example, in 1860, a group of Nogay migrants who left the Caucasus and were resettled in the Çukurova plains sent a petition to be moved higher up to the Taurus Mountains. Similarly, in the same period, several communities of nomadic pastoralists refused to comply with their permanent settlement orders and resumed their seasonal migration.[11] The rising toll of malaria during and after World War I on both the rural populations and the Ottoman army also coincided with changing geopolitics and agricultural practices. Alongside the expansion of rice cultivation, for instance, came a new strain of malaria, *Plasmodium falciparum*, probably carried by Ottoman troops returning from India and Egypt.[12]

Officials of the Turkish republic, founded in 1923, viewed marsh and swamp reclamation as material and political tools to accommodate population exchanges and migrations in new agricultural land. High-modernist projects for altering terrain and ecologies in such impassable areas as marshes, mountains, and forests were deployed to control and assimilate unruly populations that lived in these areas.[13] In 1926, the Turkish government passed the Law for Fighting Malaria.[14] Malaria control combined public health initiatives and large-scale environmental changes to eliminate mosquito-carrying areas of saturated land and engender biopolitical and demographic transformations.[15] The renewed political motivation to undertake wetland drainage all over the country began with the remaining wetlands of Ankara, the new capital city.[16]

Wetland drainage in the Turkish republic was concurrent with the mass resettlement of Muslim populations from the former Ottoman provinces and, later, of rural migrants, including peasants displaced in the construction of large dam reservoirs.[17] Beginning in the 1950s, state-led projects of wetland drainage aimed at creating large swaths of new agricultural land for land redistribution, boosted by legislative measures such as the Law of Drained Wetlands and Reclaimed Lands.[18] This law allowed landowners to appeal for drainage of state land and receive compensation for land acquired through eminent domain, and it established guidelines for redistributing newly drained land. After 1953, the Turkish DSI, an agency modeled on the US Tennessee Valley Authority, would be in charge of the hydraulic infrastructure and policies that enabled and sustained drainage. Reclaimed marshland continued to be used to resettle Muslim populations exchanged from former Ottoman countries in the Balkans, especially Greece, Albania, and Bulgaria.

Between 1950 and 1990 Turkey lost approximately half of its wetlands to varied hydraulic reengineering and transformations of environments. However, beginning in the 1960s—with malaria epidemics under control and the beginning of new processes of urban migrations that would depopulate large sections of the countryside—Turkish scientists and bureaucrats began to debate how to balance imperatives of wetland conservation, resource extraction, and national development. In doing so, they joined a conversation that had previously centered in Western Europe and the United States. The following sections focus on the rise of Turkish ornithology, then examine the emergence of ecological categories of wetland in the first half of the twentieth century, before returning to Turkey's embrace of international wetland discussions from the 1960s, ultimately leading to an institutional apparatus of technocratic wetland management.

The Rise of Turkish Ornithology

While Ottoman and Turkish officials, like those in many other empires and states, were concerned with the drainage of swamp forests and marshes, an early network of wetland conservationists was emerging. Ottoman Turkey, long a field site for European ornithologists and amateur birders, would soon become known for its widespread wetlands. Local bird experts worked as useful fieldwork allies for visiting European ornithologists and, in time, would form autonomous ornithological societies. This Turkish network of ornithologically passionate scientists, state officials, and civil-society organizers would come to embrace the new imperatives of wetland protection.

Despite widespread belief in the insalubrity of shallow water, birds—including marsh-loving species—played important roles in Ottoman and Turkish life. Birds were often central protagonists of Ottoman poetry and featured in artistic representations, religious narratives, and folk tales. Ottoman aristocrats acquired and trained birds of prey for falconry hunting and collected exotic birds in their palace aviaries. Houses and villas—also bridges, mosques, schools, and fountains—frequently included elaborate birdhouses embedded in their architecture.[19]

European scientists have largely portrayed themselves as central to the "discovery" of Turkish birds, eliding the stories of their Ottoman and Turkish collaborators. Accounts of regional ornithology usually begin with the travels of European scientists in the region, starting with the French naturalist Pierre Belon in the mid-sixteenth century,[20] and spanning to the late nineteenth-century travels of the wealthy natural history collector Charles Danford. Danford, like many ornithologists of his time, organized scientific trips to collect specimens for his collection back in London. In the 1870s Danford shipped thousands of birds acquired during his Ottoman Turkey expeditions back to England. The shipments included a valuable bald ibis from the town of Birecik, which had been shot despite the residents' resistance, as the bird held spiritual importance and was considered sacred.[21] The work of local collaborators was essential to the expansion of European ornithology in Ottoman territories. For example, in 1891 the British ornithologist Henry Dresser received a shipment of birds and eggs collected in Erzurum by the British consul, a Persian-Armenian intellectual. The specimens had been packed in sawdust, which filtered out of the crates during the journey. By the time the crates arrived in London, the birds and eggs had been crushed. The shipment contained no annotations or labels but a short list of "common" bird species observed in Erzurum. Dresser, dismayed, added the few broken specimens he was able to repair to his growing collection of taxidermic birds and eggs in London.[22]

Many middle-class European expatriates to Turkey were also keen bird collectors. For example, Thomas Robson, a British amateur ornithologist who worked in an iron manufactory near Newcastle as a clerk and mechanic, moved to Istanbul in 1862, seeking a more salubrious climate, and continued to collect and study local avifauna.[23] In his correspondence with a zoologist at the London Natural History museum, Robson described the shooting parties he occasionally took to the countryside. He complained about the great difficulties of collecting birds outside Istanbul, for roads were scarce, and the local populations rude and hostile.[24]

Alongside traveling naturalists, armchair ornithologists, and expatriate birders were a few Turkish scientists plugged into international scientific networks. Zoologist Ali Wahbi studied the ornithology of Istanbul, conducting research throughout the end of the Ottoman Empire and the founding of the Turkish republic. In a 1929 publication, as the first Turkish citizen to publish an international ornithology article, Wahbi classified the migratory and sedentary birds he observed (and captured, killed, and stored in the zoology museum) in Istanbul's lakes, rivers, forests, and seas.[25] Wahbi noted the spring and fall voyages of hoopoes, quail, turtledoves, pigeons, magpies, wheatears, and rollers. He took note of rare spottings, such as an arctic red knot he observed one day in November 1919, as thousands of cormorants hungry for fish, "a real plague for the region," blackened the sea and paraded on the waterfront at Moda.[26]

Endangered Birds and the Moral Call of Preservation

While Ottoman Turkey was emerging as a prominent site of ornithological interest for European and local birders, a separate network of European and North American ornithologists was coalescing around a shared interest in preserving waterbirds and, as a means to that end, their watery habitats.[27] These middle- and upper-class ornithologists, birders, and hunters framed the preservation of such sites as a moral duty for civilized societies. Many were motivated by a moral mandate for environmental conservation. These motivations were sometimes entangled with an understanding of Anglo-Saxon superiority, and ideas of wilderness loss were entangled with fear of racial extinction in the face of growing immigration to North America and the presence of the descendants of enslaved Africans.[28] Others were also concerned with the problem of international cooperation over the management of bird stocks—a problem that required techniques for counting birds and monitoring their population loss. Yet others brought a scientific, aesthetic, and mystical appreciation for the ecological uniqueness of environments season-ally saturated in water.[29] These heterogeneous communities of conservation-ists ultimately transformed the wetland, a category initially created around the lives of birds that were at the center of such hunting and scientific projects, into an ecological concept with a much wider outreach: a platform for inter-national scientific discussion and regulation.

In the late nineteenth century, North American bird lovers and ornitholo-gists mobilized against the growing popularity of bird hats.[30] Outraged by the increasing rate of killing and trade of both exotic and native birds, upper-class women led these efforts,[31] and they embraced the new activity of bird-watch-ing.[32] Members of the Audubon Societies in the United States organized yearly bird counts;[33] they also supported the establishment of a nationwide network of bird refuges.[34] In 1902, bird legislation in the United States centered on dif-ferential protection for various bird species and groups, with seasonal restric-tions on the hunting of select species of game birds, as well as provisions for the protection of birds that were useful predators of agricultural pests and birds that held aesthetic value or scientific importance, including for the sci-entific collection of taxidermic specimens. Other species were "injurious and therefore excluded from protection," wrote Theodore Sherman Palmer, assis-tant chief of the US Biological Survey, while also admitting that these distinc-tions were "necessarily arbitrary."[35]

In the 1920s and 1930s, scientists and elite hunters in Europe began to take notice of the new hunting regulations and "bird refuge" networks in the United States that attempted to counter the rapid decline in waterbirds.[36] In 1922, American taxidermist and biologist T. Gilbert Pearson, an adamant opponent of the millinery trade, together with French ornithologist Theo-dore Delacour, founded a lobbying group called the International Council for Bird Preservation (ICBP). Fifteen years later, the British section of the ICBP conducted a comprehensive Europe-wide survey. This International Wildfowl Inquiry (IWI) brought together sports hunters, bird-watchers, hunting-mag-azine editors, and natural scientists (many dabbled in all four occupations).

One of them, Percy Lowe, who was curator of birds at the Natural His-tory Museum in London, president of the British Ornithologists' Union, and a passionate hunter, reflected on the problems hunters and ornithologists were encountering with the decreasing numbers of European wildfowl. Bird loss was an international problem.[37] Ducks and geese that hatched in Canada, for example, were shot when they migrated across the US border "without any regard for their future welfare." Lowe also warned against the commercialization of wild species, which would inevitably lead to extinction.[38] He advanced a moral argument for establishing international agreements on bird hunting, for "no nation has any moral right to shoot these ducks without limit."[39] Like many other hunters and ornithologists of the time, Lowe connected the decline of bird populations to the transformation of their watery habitats: the drainage of marshes, swamps, and fens that destroyed the wintering sites of birds and "countless forms of life," and the lowering of the water table deriving from drainage that resulted in the "desiccation of large agricultural areas and the sweeping away by winds of the superficial soil." Lowe described the vital hydrological functions of marshes and swamps by using the mechanical metaphor of the sponge: they held up water, preventing the flooding of agricultural land.[40] In following decades, the sponge metaphor would remain a recurring image to describe the functioning of wetlands.

Meanwhile, in Turkey, a German ornithologist who escaped Third Reich Germany helped establish the country's first bird preserve. Curt Kosswig lived in exile in Istanbul between 1937 and 1955 with his wife, Lenore, a botanist. After his arrival at Istanbul University, Kosswig founded the Department of Zoology and the Natural History Museum. He and Lenore conducted research in the wetlands of Lake Manyas, and Kosswig successfully petitioned Turkish authorities to create a "bird paradise" to protect spoonbills, herons, and cormorants. In 1958, the lake would become a national park.[41] Although Kosswig did not appear concerned with the hydrological functions of the lake's wetlands, his research and conservation advocacy at Lake Manyas planted the roots for a growing network of Turkish ornithologists who would later come to embrace the new concept of the wetland.

Postwar: From Bird Refuges to Ecological Laboratories

In the post–World War II period, this rising interest in preserving waterbirds and their vanishing habitats was the seed for a sprouting and tight-knit network of organizations, ones that remain at the forefront of international wetland advocacy to this day.[42] In 1947, ICBP members created a permanent office, called the International Waterfowl Research Bureau (IWRB). This organization of scientists and hunters, headquartered at the Natural History Museum in London, would serve as an international body for coordinating research and conservation of wetlands and waterbirds.[43]

In 1954, the IWRB moved to the newly founded Biological Station of Tour du Valat in France. It also added "wetlands" to its name, becoming the International Wildfowl and Wetlands Research Bureau. In 1968 IWRB moved again, this time to the village of Slimbridge, near Gloucester, which already hosted the headquarters of the Wildfowl Trust, founded in 1946 by Sir Peter Scott.

The social worlds of postwar European conservationists were closely inter-connected. For example, Scott would later be known as a cofounder of the World Wildlife Fund, together with the eugenicist Julian Huxley, the first UNESCO director, who also founded IUCN. Scott was also a good friend of John Berry, the zoologist who edited the International Wildfowl Inquiry reports; they had studied together at Eton and Cambridge. At Slimbridge, Scott and his staff would spend days immersed in ornithological research, with the stated aim of preventing the decline of waterbirds. They used bird decoys and net guns to capture and ring migratory ducks, studied winter flocks of wild geese in the grassy salt marshes, and hand-fed their collection of captive waterfowl—four hundred birds living in pens and artificial ponds on the estate's soggy fields.[44]

At the same time that these waterbirds and wetlands research and conservation groups took on momentum, a separate network of European ornithologists was continuing to coalesce in Turkey. The German scientist Hans Kummerlöwe undertook several ornithological research trips across Turkey on a motorcycle in the 1930s and then again in the 1960s (under the new name of Kumerloeve, in an attempt to hide his Nazi past), accompanied by his friend and colleague, Günter Niethammer. Kummerlöwe and Niethammer published extensively from their research on Turkish ornithology.[45] By the end of his career, Kummerlöwe had published more than one hundred articles on Turkish avifauna and more than twenty articles coauthored with Niethammer.[46] As a member of the Nazi Party and founder of a Nazist student union, Kummerlöwe also devoted his research to studies of Aryan racial supremacy, including an "anthropological" study of Polish prisoners conducted in an Austrian concentration camp.[47] His friend Niethammer, who volunteered to be posted at Auschwitz, undertook a chilling ornithological study of the birds living in the rivers and marshes of the extermination camp.[48] The political affiliations of these two doyens of Turkish ornithology are never mentioned in regional ornithology books, which celebrate their contributions,[49] a standard reference for midcentury ornithological data.

Until the 1960s Turkish ornithological research was largely characterized by the seasonal expeditions of German, British, and Dutch scientists. With the exception of Lake Manyas National Park, there were no wetland conservation projects undertaken in the country. Turkish state actors and institutions directed their efforts toward hydraulic remakings of Turkish landscapes through the construction of large dams, irrigation systems, and drainage schemes. Yet Turkish state foresters often shared an amateur interest in ornithology. For example, during a bird-collecting expedition in 1960, a Yale ornithologist named a new subspecies of snowfinch *Montifringilla nivalis fahrettini* to acknowledge the help of a Forestry Service official, Fahrettin, who had accompanied him in his travels.[50] Yet in looking to document the country's avian wealth, researchers were less interested in talking to villagers. In 1970, Cambridge University scientists, during an ornithological field expedition to Turkey, were displeased to find that "it is almost impossible to find a wild, 'birdy' spot free of Turks, especially in the vicinity of wetland areas."[51]

Wetlands as Laboratories

In the 1960s an all-encompassing definition of the wetland emerged, becoming a shifting "boundary object" of international conservation conferences and treaties. This broad definition of the wetland became naturalized as scientists compiled international wetland inventories. It also generated new debates about ecological value—much predating the 1990s rise of the field of "ecosystem economics." Despite the geographical reach of comparative wetland surveys beyond Europe, including in Turkey, debates over wetland conservation in this period remained largely confined to Euro-American scientific circles.

In the early 1960s, the rising interest in ornithology and concern with the loss of waterbird habitats morphed with the widening field of wetland ecology.[52] Luc Hoffmann, the son of a wealthy Swiss pharmaceutical industrialist, helped make ecological notions of the wetlands relevant to the European network of ornithologists and hunters involved in the preservation of waterbird habitats. In 1954, Hoffmann founded a research station at Tour du Valat, an estate he had acquired in the Camargue, on the Rhone Delta in southern France—a region where he had conducted ornithological research for his doctorate.[53] In the estate's expansive wetlands, brackish marshes, commercial saltpans, and agricultural fields, the Tour du Valat institute hosted research activities in avian ecology, ornithology, wetland ecology, and conservation.

Speaking at the International Congress of Zoology in 1963, Hoffmann argued that wetlands, characterized by "the most diverse habitats" shaped by cyclical flows of wet and dry, amphibious life-forms, and the possibility of regulating hydraulic dynamics, were ideal sites to test ecological hypotheses.[54] He advocated for moving beyond making lists of birds and wetlands. Rather, these inventories and collections were merely "an instrument to be used for ecological research."[55] Drawing on novel understandings of ecosystems as energy transactions in an age of nuclear research,[56] Hoffmann postulated that wetland research should be based on questions about ecosystem structure, relationships, and energy exchanges among its different components. Classification of wetland fauna according to "trophic levels" would reveal energy pyramids: the "raw structure of the biotic community."[57] Hoffman characterized wetlands as controllable and engineered open-air laboratories.[58] Wetland managers would be able to control water flows so that scientists could manipulate wetlands to test their hypotheses. They could, for example, intervene in processes of ecological succession to favor the habitats of certain species. The laboratorization of the wetland would involve ecologists burning or cutting trees and plants, increasing the productivity of wetlands with artificial fertilizers or with cattle, draining waters to eliminate large fish, and even replacing fish species.[59]

Meanwhile, in France, under Luc Hoffmann's leadership, the Biological Station of Tour du Valat had undertaken an international program to promote wetland conservation science, starting with a focus on Mediterranean wetlands. Hoffmann named it Project MAR—"mar" being the first three letters for wetland in English (marsh), Spanish (*marisma*) French (*marecages*), and Italian (*maremma*).[60] Project MAR brought together the IUCN, IUBP, and IWRB (at the time these were the major international environmental organizations in Europe) and was supported by UNESCO funding. In fact, Hoffmann was personally involved with all these institutions: he was vice president for the IUCN between 1960 and 1969 and directed the IWRB until 1968. He also helped cofound the WWF with Scott, Huxley, and others.[61]

The first Project MAR conference was held in November 1962 near the Tour du Valat research station, at Les Saintes-Maries-de-la Mer, attended by eighty scientists from government offices, universities, and environmental organizations, mostly from France, the Netherlands, the United Kingdom, and the United States. At the conference, they tackled the economic, scientific, and moral reasons for conservation; evaluated wetland definition criteria; and agreed on the need for international efforts. Later, conference delegates would remember the sumptuous meal that Madame Hoffmann served at the family home at the end of the conference or the lone spotted eagle fluttering in the marshes of Tour du Valat amid thousands of flamingos.[62] But for Luc Hoffman and others, the conference would be remembered as ground zero for the creation of an international network of wetland conservation. Debating various scientific and political arguments for wetland conservation entailed detailing what wetlands were and discussing multiple and contrasting approaches to conservation. For example, US ecologist Eugene Odum emphasized that marshes, flats, creeks, and bays in an estuary should be considered a single ecosystem with a natural capacity for production—an equilibrium threatened by marshes' conversion to monocrop agriculture. Representatives from US Fish and Wildlife decanted the "value of wetlands to modern society,"[63] while a Dutch forester focused on the infrastructure for controlling wetlands' water levels.[64]

Looming large in the discussions were the effects of ongoing large-scale wetland drainage and reclamation projects that, as most participants made a point of noting, had already destroyed wetlands all over the world. For some, wetlands were valuable as ecological habitats for waterbirds or in their own right; for others, wetlands guaranteed the continued supply of economic resources. In his conference remarks Sir Peter Scott critiqued the purely economic arguments for wetland protection employed by many scientists: the destruction of the marshes, he declared, was "an unthinkable crime against civilization." The "mature civilizations" of Europe and North America should privilege scientific research over economic concerns.[65] Animated discussions among the participants ensued about whether the reasons for conservation should be economic, aesthetic, or motivated by teaching and research goals and how to convince governments to shift their economic policies from the drainage of marshes and wetlands to the establishment of a network of wetland reserves.[66]

The final recommendations of the MAR program created a "boundary object" that encompassed all these competing arguments.[67] Wetlands were "habitats of the greatest importance and interest to humanity" because of "their natural biological productivity and by their educational, scientific, cultural, economic and recreational values." Governments should "make provisions for wetland reserves in all national and development plans" to counter the destruction of wetlands "as a result of drainage, industrialization and

other changes in land use." The establishment of national reserves would protect some of the remaining wetlands from environmental changes wrought
by industrialization and agricultural expansion. The conference proceedings
urged increased collaboration between scientists and governments as a solu-
tion to the environmental problems resulting from development. To this end,
they urged collaborating scientists to compile a comprehensive list of European and North African wetlands "of international importance" to be made
available to conservationists and national planners to counter the rapid drainage of those regions. This list would be foundational to a future international
wetland convention.[68]

Thus, Project MAR transformed wetland conservation into a global technocratic palliative for the destruction wrought by economic and technological development worldwide. Project MAR set out with a series of ambitious
interconnected goals. It would prepare a statement "about the importance of
marshes and wetlands to modern mankind" in a time of rapid environmental
degradation and assemble data on wetland conservation, management, restoration, and wildlife to "make this information known and available to all
those in a position to take action to advance the conservation of wetlands."
Project MAR scientists would also provide governments with technical assistance for establishing wetland reserves. Central to this effort was a worldwide
inventory of wetlands of "international importance," starting in Europe,
North and West Africa, and Russia.[69]

Liquid Assets

As wetlands became ecological objects, scientists also sought to popularize
their value among ordinary citizens. In 1964, wildfowl hunter and duck counter Georges Atkinson-Willes, a veteran amputee who worked for the Wildfowl Trust at Slimbridge, wrote a short book titled *Liquid Assets*. The book
was commissioned by the IWRB, where Atkinson-Wilkes worked as the
coordinator for hundreds of volunteers participating in yearly "duck counts,"
and UNESCO paid for printing and distribution.[70] The intended readership
for *Liquid Assets* was "everyone whose work has a bearing on the future of
marshes and wetlands in Europe"—wetlands that were rapidly vanishing.
Atkinson-Wilkes described wetlands as a natural resource, like farmland
or forests, valuable and yet increasingly scarce: "every effort must be made
to preserve what still remains."[71] *Liquid Assets* also popularized a broad and

all-encompassing definition of wetland, which would later be adopted in the international wetland treaty:

> By wetlands we mean all areas of marsh and all stretches of water less than twenty feet (six meters) deep, whether fresh or salt, temporary or permanent, static or flowing. Important categories include estuaries and coastal shallows, brackish and saline lagoons, natural and artificial lakes, complexes of small ponds or pot-holes, reservoir and gravel-pits, rivers, swamps and flood-meadows.[72]

These varied habitats supported "a vast and specialized range of plant and animal life, the full value of which is only now being realized."[73]

Following the 1962 Project MAR conference at Tour du Valat, and a year after the publication of *Liquid Assets*, the Project MAR team completed its first inventory of wetlands in Europe and North Africa, drawing from data obtained during scientific expeditions as well as through postal correspondence with over five hundred European scientists. Despite its stated aspirations to international cooperation, the Project MAR inventory remained rooted in European perspectives. Data collection beyond Northern European locales was scant and produced with little local input. For example, although the MAR report described Turkey as a country of "extreme ornithological importance," the authors conceded that they could obtain hardly any information "to compile a comparative list of sites of international importance."[74]

The Project MAR report made apparent the problem of commensurability arising in the effort to create a global wetland category. What criteria would be used to define and classify wetlands and to determine their "international importance?" Even as the main definition of wetlands, such as the one advertised in the *Liquid Assets* pamphlet, was all-encompassing, subclassifications abounded. Wetlands were divided into eight ecological categories, defined hydrologically, according to the quality, location, and flow of their waters: coastal waters; shallow coastal lagoons; coastal marshes; shallow inland salt, brackish, or alkaline water; shallow, static inland freshwater; inland freshwater mineral-marshes; and peatland. Each of these was divided into subcategories, which drew on more heterogeneous features.

For an example of how these classifications worked, consider the category "inland freshwater mineral-marshes." These were defined as waterlogged sites, permanently or temporarily saturated, with emergent vegetation. The soil would be mineral or inorganic, without peat. The marshes could be isolated or interconnected in marshy belts around deeper lakes, could be located at different altitudes (uplands or highlands), and could feature trees, thickets, or meadows. When the vegetation was permanently waterlogged, these ecologies would be classified as a swamp; if they were only temporarily waterlogged, they would be a "moist habitat."[75]

Birds remained proxies for the importance and the health of wetlands.[76] For the Project MAR scientists, this was a pragmatic choice: they already had "a considerable amount of information on the ecology of birds in relation to wetlands." They also posited that bird migration routes—spanning regions, countries, and continents—generated more problems of international coordination than any other animals or plants.[77] Not all birds were of equal importance, however, and Anseriformes (an order that includes water-loving birds such as the horned screamer, the magpie goose, the tufted duck, and the mallard) received a "special emphasis" in the making of wetland inventories, "mainly because of their obvious importance" and also, more pragmatically, because there were more data available on them than on other orders of birds.[78] The broad definition of the wetland made it difficult to demarcate where the hydrological boundaries of wetlands lay, and the MAR report conceded that indeed there was no objective definition of wetland. Birds could again be invoked as a proxy to help find the elusive boundaries. As many species of birds feed in one area, rest in another, and breed in yet another environment, all these habitats should be protected, MAR scientists urged.[79] In subsequent decades, other global environmental objects—biodiversity, water, and climate—would come to animate wetland conservation as much as bird preservation did in the early decades of the century.

The concerns of European and American hunters and scientists over decreasing numbers of waterbirds, resulting from both an intensification in hunting practices and drainage and terraforming, shaped and stabilized meanings of wetlands. Defenders of reclamation relied on the assumption that drainage destroyed worthless and unutilized marshes to obtain valuable agricultural land and improve public health, advancing national development. Starting in the 1940s, however, wetland preservationists had adopted the same utilitarian arguments of their opponents, demonstrating the inefficiency of drainage schemes and the loss of increasingly scarce fishing and game resources.[80] In the 1960s, with publications such as *Liquid Assets*, wetland advocates began to leverage scientific studies demonstrating the multiple functions of the wetland—for example, for flood control, pollution remedia-tion, food production, and recreational activities.[81]

Stabilizing International Cooperation: The Wetlands of Turkey

Throughout the 1960s, Project MAR, the IWRB, and the IUCN organized a series of international conferences on the topic of waterbird and wetland conservation.[82] In these conferences, scientists and advocates sought to craft an international convention for the conservation of wetlands. In 1967, Turkey became directly involved in the conversation as the country's officials hosted the "Technical Meeting on Wetland Conservation" that opened this chapter. The conference, funded by the IUCN's Commission on Ecology, focused on wetlands to "promote an ecological approach to the conservation problems of the Near and Middle East Region."[83] The meeting was held at universities in Istanbul and Ankara and had forty participants and seven observers—includ-ing Turkish state officials, scientists, and members of the IUCN, ICBP, IWRB, the FAO, the International Biological Programme, the Council of Europe, and the Wildfowl Trust. Over the course of a week, participants discussed prob-lems of soil erosion and water conservation, Turkish ornithology, and state projects of large-scale wetland drainage for agricultural use and presented reports of their field visits to several wetland areas in the country. At the end of the conference, several delegates established a foreign and a national sec-tion of the Turkish Ornithological Society.[84]

A young ornithologist, Tansu Gürpinar, led the efforts to start the Turk-ish branch of the ornithological society. Trained in the natural sciences at Ankara University, in 1966 Gürpinar left his job as a state geologist to join the National Parks Department, then a subdivision of the Forestry Bureau in the Ministry of Agriculture, and was posted at Lake Manyas.[85] At the 1967 wetland conference, Gürpinar presented a paper on Curt Kosswig's Lake Manyas preserve, which by then had become Turkey's first national park. The lake was also the country's only "Bird Paradise" (*Kuş Cenneti*), a Turk-ish appropriation of an older English and German term. Just months before the conference, the Turkish government had tasked the Council of Europe with sending a consultant to help with management and administration of the wetland preserve. David Lea, ornithologist at the Royal Society for the Protection of Birds, recommended extending the boundaries of the park, building bird-watching facilities for visitors, trimming the phragmites, flooding nearby fields to create feeding grounds for grey herons, planting trees for squacco heron and glossy ibis nesting, pruning others to attract Dalmatian pelicans, and removing some of the willows planted in an area favored by ducks. While Lea focused on ecological landscaping to curate bird habitats, Gürpinar was growing attuned to local responses. To succeed, the conservation officers should "make a propaganda effort in neighboring villages in order to encourage more care for the breeding colonies of birds" and monitor the rapid expansion of agricultural production in the area.[86] The Lake Manyas preserve was established in a thickly agrarian environment, like most other wetland conservation sites in Turkey—near marshes that had been drained to resettle migrants and refugees. The questions of villagers' participation in or resistance to conservation and of agricultural transformations would be central to Turkish wetland management for decades to come.

Three years after the Turkish wetland conference, Gürpinar, who had then been appointed director of the Lake Manyas National Park, traveled to Iran to represent Turkey at a large international conference for conservation of wetlands. Between January 30 and February 3, 1971, eighteen delegates met in the seaside Iranian town of Ramsar, with the main purpose of preparing the text of an international wetland conservation agreement, called the Ramsar Convention.

The Final Act of the Ramsar treaty had twelve articles. The first was a technical definition of wetlands. The second defined the duty of each contracting party to designate sites to add to a list of wetlands of international importance "on account of their international significance in terms of ecology, botany, zoology, limnology or hydrology." The third article stipulated that "the Contracting Parties shall formulate and implement their planning so as to promote the conservation of the wetlands included in the List, and as far as possible the wise use of wetlands in their territory." The fourth article indicated the duty of each party to establish nature reserves on wetlands, exchange wetland research data, increase waterfowl populations, and train wetland experts and managers. The remaining articles established a triennial conference of the parties and defined the institutional structure of the convention, including the modality of accession.[87] UNESCO opened the convention for accession in 1972, after sending copies of the Final Act to the governments of 121 countries, and the convention was formally ratified in 1975, as seven European countries signed it. Three years after having successfully organized the

1971 wetland convention at Ramsar, the IWRB replaced the word "wildfowl" in its name with "waterfowl." A Ramsar Convention office was created separately and headquartered with the IUCN, the UNESCO organization that had helped fund IWRB wetland conservation efforts since the early 1960s.[88]

Although Turkey would not sign onto the convention until the mid-1990s, wetland conservation had now become a national concern, generating civil-society activity and interest and countless debates. For example, in 1975, Gürpinar founded an environmental association, the Society for the Conservation of Nature (Doğal Hayat Koruma Derneği [DHKD]). Ornithologists who were affiliated with the society in the 1980s found employment in Izmir, Istanbul, Samsun, Ankara, and other regions of Turkey and remained involved with wetland conservation, whether as researchers or as founders and members of environmental NGOs. In 1980, Ramsar held its first "Conference of the Parties" in Sardinia, Italy, with representatives from the twenty-one signatory countries; from then on, it was held triennially. Initially, Ramsar membership grew rapidly among countries in Europe and in the Americas. Only in the late 1980s and early 1990s did Asian and African countries, Turkey among them, begin to join the convention.[89]

Turkish citizens' questioning the effects of large-scale drainage politics had already begun in the 1960s, but after the 1980s it became a more widespread matter of concern, and many environmental NGOs took up the task of wetland protection. Several environmental organizations were founded in the 1980s and 1990s in Ankara and Istanbul. For example, the Environmental Problems Foundation of Turkey (Türkiye Sorunları Çevre Vakfı [TÇV]), founded in 1978, lobbied for inserting environmental protection in the Turkish constitution. TÇV also published the first nationwide survey of Turkish wetlands. "Wetlands are areas to which we pay little attention in our everyday life," wrote the report's authors. "We often disparage them by referring to them as swamps. But, with their special characteristics and the living creatures to which they offer refuge, they possess an enormous importance in sustaining the ecological balance." Turkey, the authors boasted, "possesses the largest expanse of wetlands among the European and Middle Eastern countries [after the Soviet Union]. Around 250 of the 400 species of birds in Turkey are migratory; many find refuge in the country's wetlands." The authors urged Turkey to sign the Ramsar Convention, for "the protection of wetlands and the living creatures that inhabit them constitutes an inseparable part of activities to protect the environment and preserve the ecological balance in Turkey."[90]

One day in October 1988, a Turkish high schooler with a passion for wetland ecology and ornithology accompanied Turkey's general/president Kenan Evren on a visit to the Sultansazlığı marshes near the city of Kayseri. General Evren had spent the previous eight years reducing civic liberties and human rights; his military rule and presidency were characterized by mass political arrests, executions, news blackouts, curfews, extrajudicial killings, prison torture, and creation of a repressive political system. But on that day, the young ornithologist, Uygar Özesmi, asked the general/president to intervene on behalf of birds' livelihoods. A long-planned irrigation and drainage project engineered by the DSI was going to desiccate the marshes.[91] In the shallow wetland lakes at the edge of growing agricultural areas and reed beds, tens of thousands of birds of almost three hundred species thrived—flamingos, herons, geese, ibises, spoonbills, ducks, cranes, waders—alongside reptiles, amphibians, fish, mammals, and plants. The DSI had agreed to maintain a certain level of water in the marshes, now supported by water from dammed lakes upstream, but conservationists worried about the devastating effects of agricultural runoff flowing into the wetland lakes, as well as the irrigation-induced rapid fluctuation in water level and the use of the area for reed cutting, grazing, egg collecting, and hunting.[92]

A year after the dictator's visit to the wetland, an agricultural engineer, Osman Erdem, was hired as a civil servant at the Ministry of Environment. Over decades of work at the ministry, Erdem authored countless reports and books on wetland conservation and birds, including the Turkish publication *The Importance of Wetlands* in 1994, published in English translation a year later as *Turkey's Bird Paradises*.[93] The book marked Turkey's joining the Ramsar Convention and the legislative efforts to create the legal and institutional system to manage and preserve Turkey's wetlands "of international importance."[94] In a preface to *Turkey's Bird Paradises*, Turkey's minister of environment Riza Akçalı balanced the "necessity of development" with the need to protect wetland ecosystems "at all costs." He exhorted his fellow citizens that "the pace and volume of development should be such as to ensure continuity, for in the long term the continuity of the economic system is primarily dependent on the preservation and development of the environmental assets which feed that economic system."[95] Erdem negatively assessed the drainage projects that the DSI had conducted in the 1950s, which had led to the drying out and disappearance of Lakes Amik, Emen, Avlan, Suğla, Kestel Efteni, and Simav and the Aynaz and Karasaz swamps, among many other sites. With the disappearance of wetlands, he elaborated, water regimes had shifted, wildlife had disappeared, and local climates had changed. Alongside drainage canals, flows of fertilizers, pesticides, and industrial and urban run-off had become a new potent threat to wetlands. Reclaimed land had proven to be much less productive than agricultural engineers and hydrologists expected, Erdem explained, echoing similar utilitarian arguments advanced by North American conservationists in the 1930s.[96]

Proliferations

In 1993, the authors of an Indonesian training manual for the Asian Wetland Bureau conceded that the term "wetland" meant different things to different people, and about fifty scientific definitions coexisted. Some definitions were narrow, for example, based on the presence of plants that thrive in both terrestrial and aquatic environments. Broader definitions of the wetland, like the one in *Liquid Assets*, encompassed reef flats and seagrass beds, mudflats, mangroves, estuaries, rivers, freshwater and saline marshes, swamp forests, and lakes. These more general definitions operated at the level of the water basin, which also corresponded to the scale of land planning.[97] The broad definition of the wetland, then, held sway partly because it could more easily be translated into land-planning schemes and partly because it could more easily become a global category, one that traveled and generated international collaboration, national legislation, and scientific projects.

In the mid-1990s Turkey increased the number of wetlands classified as "wetlands of international importance" and designated eight Ramsar sites. In 2002, the Ministry of Forestry and Environment issued a legislative act, the Regulation for the Protection of Wetlands.[98] It also instituted a National Wetlands Commission, constituted of representatives across different ministries and university experts, and a Wetlands Department. The ministry also set up wetlands commissions operating at provincial and local levels, prepared a series of five-year plans for National Wetlands Strategy, and commissioned wetland management plans for all Ramsar wetlands. Alongside this mushrooming governmental wetland apparatus, university scientists and NGO staff crafted their own projects of wetland research and advocacy. In 2009, several universities and environmental NGOs organized the first biannual National Wetland Conference, with governmental support and oversight. In 2014, revisions to wetland legislation introduced different categories of local and national wetlands. The 2014 wetland law also instituted new procedures for opening wetland areas to infrastructural development. These legislative changes were harshly criticized by environmental NGOs, especially for facilitating the process of eminent domain within the preservation zones, and remained at the heart of grassroots contestation of infrastructure development within wetland areas, as the following chapters show.

Gürpinar's association, DHKD, remained among the most active in running conservation projects with international funding. The association collaborated with the Ministry of Environment to write the first wetland management plan in the country, in the Göksu Delta (one of Turkey's first Ramsar sites). In the early 1990s it undertook studies for integrated management and, starting in the mid-1990s, a project on integrated development and conservation.[99] In 1998, DHKD successfully lobbied for the inclusion of four more wetland sites to the Ramsar list, including the Gediz and the Kızılırmak Deltas. In 2001, members of DHKD split: one group formed an independent association, Doğa Derneği, which became the national partner of Bird Life International (the successor of the International Council for Bird Preservation), while DHDK became Turkey's WWF. All these associations were scientifically oriented, with a membership of university scientists and state officials, but also included many civil-society organizers.

Until the end of the 1990s, Turkish conversations about wetland conservation remained confined to the scientific and aesthetic interests of middle-class and educated professionals in the country's metropolises. By the time of my research, wetlands had been enfolded into more widespread grassroots mobilization and popular concerns over environmental sovereignty, biodiversity, ecological justice, and the livelihood rights of nonhuman animals and plants.

Conclusion

The transformation of Ottoman landscapes had been a vision of agricultural civilization. Marsh drainage in the first decades after the founding of the Turkish republic addressed public health concerns and the problem of resettling hundreds of thousands of refugees, exchanged populations, and internal migrants escaping rural poverty. The drainage of water-saturated environments in the early twentieth century led to the rise of a loose international network of wetland conservationists, joined by a shared interest in supporting the watery habitats of birds. These interests centered on secular and elite practices of hunting, birding, and ornithology—following the expanding boundaries of imperial ecologies and scientific explorations concurrent with colonial conquest. In the 1950s, scientists drew on idioms of ecology to create wetlands as an ecosystem form. Wetlands became all-encompassing boundary objects for international cooperation over environmental conservation. Turkish birders and ornithologists, often cast as secondary characters in the tales of European scientists, paved the way in embracing the category "wetland," now framed around idioms of loss. They did not simply absorb it but made it their own, articulating concerns over the ecological destruction wrought by national development.

In a 2008 article in a Turkish photography magazine, Tansu Gürpinar described the feeling of approaching a wetland, camera in hand. "You are near a wetland when you feel more light around you, your exposure increases, you feel a sense of relief; you find yourselves in a place in which bird calls are louder, and you can hear water gurgling, waves, the tumbling of waterfalls."[100] Wetlands were no longer only areas for the reproduction of bird stocks, sites of open-air laboratories, or functional to the conservation of natural resources. For Gürpinar and many others, the wetland had become a place of sensorial attunement, a place of synesthetic aesthetic contemplation that one might visit with a camera, seeking to capture the play of refracted light or, through the use of telescopic lens, to portray a close encounter with wetland life-forms.

At the same time, wetland landscapes engendered a sense of loss, one closely entangled with national politics. "The loss of wetlands in Turkey continues," Gürpinar wrote. "It is apparent, from the loss of wetlands and the problems of water scarcity experienced in recent years in many parts of the country, that the institutions in charge of and responsible for water management have made serious mistakes."[101] Social scientists, in Turkey and beyond, have often focused on the structural processes that led to wetland conservation failure, providing an answer to the overdetermined question of environmental degradation. More rarely have they addressed how people come to understand lived environments undergoing different kinds of destruction and precarity and how they assess ecological change in the context of their lifeways.[102]

In many scholarly accounts of Turkish environmental policy, a paternalistic state characterized by top-down governance, alongside citizens prioritizing questions of economic development and identity politics over environmental concern, are to blame for Turkey's "biodiversity crisis."[103] In this way, social scientists have reproduced a modernist narrative, whereby Turkish environmentalism is cast as a global import, a modern standard of governance that Turkey has failed to adopt or, alternatively, a proxy for other political concerns. I propose a different narrative, one that centers on how Turkish bureaucrats, scientists, and environmental advocates participated in wetland conversations, building on the legacy of decades of ornithological interest. These scientists, bureaucrats, and environmentalists worked to make sense of changing ecologies and crafted new institutions and politics around wetland ecosystems. Sometimes, these transformations paralleled state visions of hydrological control and expert practices that marginalized rural residents living and working in the wetland areas.

In the twenty-first century, Turkish wetlands have become semantic and material nodes in the transnational movements of birds and scientists, in shifting and precarious reconfigurations of water and land, and in unstable flows of environmental governance and capital. Turkey has long been imagined by writers, artists, and diplomats as a crossroads—alternatively a bridge, a portal, or a barrier between Asia, Europe, and the Middle East; between East and West; and between tradition and modernity. Such dichotomies continue to characterize much scholarly work as well as popular writing. In contrast, Turkish governmental and NGO reports on the country's environments characterize Turkey as a network of transnational migration of species, its wetlands constituting passage points connecting distant continents, regions, and ecologies through the seasonal long-distance voyages of birds (the migratory movements of fish, microorganisms, plants, and people continue to receive much less attention in the Turkish wetland literature). Wetlands reveal the fluidity of cultural and ecological boundaries, networks, and dichotomies.

After their invention as environmental categories and foci for international and national science and management, wetlands have not been stabilized: they remain in flow both physically and discursively and are very much contested. They have become relevant to different communities—of farmers, fishermen, activists, scientists, and bureaucrats—as sites for imagining and constructing new multispecies relations in particular materialities: moral ecologies of the wetland. The following chapters demonstrate how an ethnography of wetland science, place making, work, and care moves away from predetermined questions of environmental degradation to explore the situated and contentious processes through which Turkish wetlands are produced as sites of moral ecology.

2 Sediments

Wetland Materialities in Izmir Bay

In August 2013, I moved to Izmir, a coastal metropolis with over four million human residents that featured one of Turkey's largest remaining wetlands. Standing on the sand spit of Homa Lagoon, one of the many shifting edges of the Gediz Delta wetlands, one can observe the city of Izmir extending over a long bay on the Aegean Sea, surrounded by gentle hills covered in scrub vegetation and forested mountains crisscrossed by steep fire-prevention trails, winding roads, and creeks. The mountains, surrounded by villages of concrete houses in gray, yellow, and pink, are a popular weekend destination for middle-class hikers and urban families going out for a typical "village breakfast." In the past three decades, the hills and plains surrounding the city of Izmir have rapidly been turned into new neighborhoods of tall apartment buildings designed to accommodate urban migrants—a visible mark of the speculative construction boom in the Turkish economy. Where the urban and industrial sprawl ends, the hills feature olive groves, fields, and small agricultural villages. The lower plains are a patchwork of intensive agricultural fields, industrial districts, and new suburban residential complexes for the growing middle class. Izmir's wetlands, forests, and hills are home to 342 species of birds.

I lived with a small group of Turkish academics and professionals in a two-story yellow house in Izmir's city center, a five-minute walk from the seaside. In the early twentieth century, the house had been an Italian family mansion—the neighbors told me. More recently, before my friends rented the

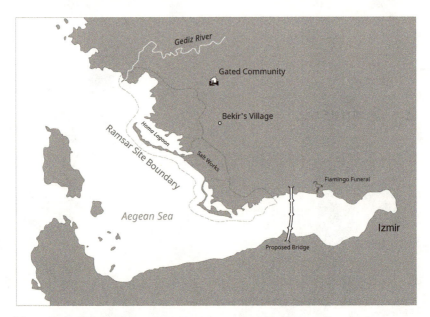

Map 2. Gediz Delta, showing the main sites mentioned in the text. Drawing by Benjamin Siegel.

decrepit building and restored it into a quasi-functional home, it had been a beauty salon, a brothel, or occasionally both, according to the carpenter, our next-door neighbor. A short walk from the yellow house was the ferry-boat station of Alsancak. The ferry dock is on a long seaside promenade constructed in the late nineteenth century, replacing hundreds of private wooden docks.[1] During the hot summer nights, I would convene with friends on the promenade's arid grass, cooling off in the sea breeze and chatting while sipping soda water and beer and snacking on pumpkin seeds, surrounded by thousands of youngsters and families doing the same. When I was not undertaking the hour-long travel to the Gediz Delta wetlands, I walked or rode my bike on the promenade on my way to the National Library, to the market in Konak, or to visit friends who lived on the southern edge of the bay. I would often sit on the promenade's walls to read a novel or ponder over my scribbles and field notes from the previous days' research.

Sitting on the concrete seawall, splashed by the waves and soaked in the acrid smells of petrol, fish, sewage, and beer, I would join a motley crowd of Roma future-tellers, urban fishermen, white-collar workers, couples, street

musicians, seagulls, university students, Syrian refugees, municipal garden-
ers and cleaning staff, feral dogs, and joggers. From dawn until late at night,
municipal ferryboats shuttled residents across the bay, while large container
freighters and cruise boats docked at the commercial port nearby. Fleets of
fishing dinghies left the fishermen's ports early in the morning to catch fish,
squid, and mussels in the bay. In the afternoons, after selling their catch at
the cooperative, fishermen often assembled at the port, tinkering with small
repairs on their boats, making tea, or cooking a late lunch on a gas stove.

From the docks of Alsancak, I could hardly see the other side of the bay,
obscured by the haze. On the rare clear and crisp days, one could see, behind
a neighborhood of high-rise apartments, seaside villas at the edge of the wetland conservation area, the municipal wastewater treatment station, and farther along the coast, tall and white mounds of salt. There, on the northern
edge of Izmir was the delta of the Gediz River, divided between the municipalities of Foça, Menemen, and Çiğli. The expansive agricultural fields of
the rural delta are dotted with sprawling gated communities, many of them
built in the last decade, and more are under construction during my fieldwork. Near the coast are a few small industrial areas and more tanneries and
metal industries upstream on the Gediz River. In Izmir Bay, the water is shallower, warm. But in the middle of the bay, deeper channels have been dredged
to allow the passage of large container ships. The coastal area of the Gediz
Delta—an ecology of lagoons, reed beds, fresh- and saltwater marshes, salt
meadows, wet meadows, and Mediterranean shrubland—is demarcated as a
wetland conservation area.

In the wake of the international rise of wetland conservation detailed
in Chapter 1, the ecological category "wetland" has become salient to many
different social groups in Turkey—from environmental NGO scientists to
fishermen, ornithologists, university and government scientists, wage laborers, bureaucrats, farmers, elected politicians, and middle-class residents—as
a contested political, cultural, and moral undertaking. This chapter argues
that everyday contestations over wetland conservation are not simply about
the ethical values of preservation, development, and livelihood, but they also
implicate material remakings of land, water, and infrastructure. Following
scientists at work through muddy estuaries, field stations, university offices,
and conferences and in their homes, this chapter frames how competing ideas
of wetland habitats and livelihood are produced, contested, and remodeled
through social and political processes.

Moral ecologies of the wetland, in this chapter, are ethically informed assessments of the materiality of hydrological and ecological processes in the wetlands' built environments. These are sedimented with historical legacies of past infrastructural and environmental transformations. Turkish scientists, NGO staff, and bureaucrats anchored their political visions and histories in the marsh while rebuilding the wetland as a habitat for endangered species. I track three ways they manage materiality as moral ecologies: building wildlife habitats for select endangered species of animals and plants, harnessing the ecosystem functions of the wetland, and inviting upper- and middle-class access to wetland ecologies as sites of leisure, recreation, and learning—at the exclusion of rural and working-class livelihoods. In the process, different communities of wetland experts worked to naturalize particular material states of the wetland. Wetland naturalizations sedimented on one another, sometimes reinforcing each other, sometimes sweeping each other away, sometimes working as patterned mosaic sedimentations. This rebuilding is a political project that harnesses and redirects socioecological dynamics for specific visions of moral order and institutional control.

Anthropologists have sometimes taken the materiality of water as a starting point, as a set of hydrological and biochemical processes that shape social practice and cultural meaning,[2] even when they can be constrained and redirected.[3] However, experts, managers, and scientists describe wetlands as constituted by specific kinds of soils, water, minerals, and organisms, governed and constrained by specific understandings of wetland quality and desirability. To materialize the wetland through invocations of scientific expertise is a sociopolitical outcome.[4] Wetlands' watery mud, sediments, and different mixtures of water, rocks, soils, rubble, plants, and bacteria also have complex social lives. At stake in contestations about different kinds of materiality are moral understandings of place and ecology. So paying attention to sediments highlights the things that water flows carry and the infrastructural politics of conservation.[5] Sediments also are residual histories of nature making, amid histories of ethnonationalist violence and displacement. The wetland is rife with political histories left in suspense, deposited in uneven layers and in unexpected forms and never fully set.

Scientists, city officials, and NGO workers in Izmir made the Gediz Delta wetland through contrasting claims of scientific expertise as well as through their everyday practices of ecological care. Research and management projects, run by universities, international research centers, ministerial offices, and NGOs, have constituted the Gediz Delta as an important node in the international networks of wetland preservation.[6] In their everyday life and work in the delta, wetland experts also leverage their research activities to craft and advance moral claims about the value of specific material and infrastructural forms of wetland ecology. I expand the anthropological concept of moral ecology, which anthropologists generally use to characterize indigenous cosmologies or wage workers and small holders' resistance against agrocapitalism and land expropriation[7]—to explain what is at stake in the desire to physically remake ecological relations and the materials and infrastructures that sustain them. This desire cannot simply be accounted for by the contrast between nonmeasurable ethical values and economic valuations of ecologies.[8] Moral economies, in classic anthropological use of the term, are notions of mutual obligations in market and nonmarket settings, reciprocal exchanges that are perceived to be mutually beneficial and embedded in ongoing social relationships. Similarly, moral *ecologies* are ethical evaluations of relationships between humans and plants, animals, water, infrastructure, soil, and other nonhumans.

Moral ecologies need not produce liberatory outcomes. Scientists' valuation of plant and animal livelihoods and of desirable water flows sometimes results in the expropriation of wetland ecologies from farmers, fishermen, and working-class wetland dwellers. The displacement of unwanted people, plants, animals, microorganisms, and even different kinds of sediments and waters is not unintentional but designed into and necessary to many forms of wetland care. Thinking with moral ecology, then, helps draw attention to the violence of care as well as to the deeply felt sense of urgency of wetland advocates, managers, and scientists. One's specific claims of what wetland materiality should be overrides others' visions and care for the wetland. In this sense, contestations over wetlands reveal power differentials while also opening novel possibilities—sometimes enabled by the seemingly apolitical nature of wetland governance—for crafting moral ecologies that defy existing hierarchies.

The Flamingo Nesting Island

One rainy morning in early October 2013, I sat on a table on the ground floor of the wetland management building in the lower Gediz Delta with a small team of scientists from the French Tour du Valat research institute. We sipped

cups of hot, strong tea, waiting for the drizzle to cease. Standing outside
the building, in the chilly autumn rain, one could see the overlapping and
entangled infrastructural ecologies of the wetland. These were the human-
built environments that produced "nature" at the edge of agricultural fields
and industrial areas The salt marshes in the decommissioned saltpans turned
into an expanse of reeds, fed by the nutrient-rich flow of irrigation water that
the wetland management agency pumped into the wetlands each spring. I
conjured the different kinds of waters, soils, sediments, and living organisms
that formed the delta—from the calm waters of Homa Lagoon to the large
drainage canal, the sparsely vegetated hills of Üç Tepeler rising from the salt
marshes, and the white, long piles of salt laid out by conveyer belts and seasonal workers in the Çamaltı Saltworks.

Anthropologists understand water to be multiple, materializing social,
cultural, and biophysical relations.[9] Some have argued that the material qualities of water and its flows invite the social meanings ascribed to it.[10] Others
assert that the existence of water as a knowable and measurable object is a
technoscientific abstraction.[11] Water flows provide rich metaphors for social
theory—for global connections and fluid social relations.[12] And wetlands'
waters are often imagined to be places of dynamic flows contraposed with
unmoving land, sites that resist our symbolic and material efforts to define
and stabilize them, and dichotomies of nature/culture.[13] But water moves in
complicated ways, anthropologist Franz Krause reminds us, "in rhythms of
varying intensity, tempo, and direction negotiated by human labor, infrastructure, the weather, and the river bed."[14] In the Gediz Delta, conversations
regarding wetland management were characterized by uncertainty and disagreement about these rhythms and infrastructures.

The Gediz Delta is constituted not only by the ways in which water flows
but also by what it carries. Anthropologist Tanya Richardson has suggested
that paying attention to the sediments that move in water, alongside toxins,
or algae blooms, might move us beyond the opposition between moving flows
and static terra. She also proposed to consider the hydrological and infrastructural outcomes of political-economic processes that have led people away
from water-based livelihoods.[15] Material remakings of ecology, through the
entangled work of infrastructural interventions and scientific claims, are central to Turkish and European natural scientists crafting wetland conservation
as practices of moral world making.

Inside the Gediz Delta wetlands management building, next to the caf-eteria, in an empty conference room, bright computer screens live-streamed thousands of greater flamingos (*Phoenicopterus roseus*) swarming in the shal-low waters of the commercial saltpans. The previous spring, a few thousand flamingos had hatched in the small volcano-shaped nests that wetland staff had constructed on top of an artificial island, hoping to attract the flamingos' seasonal nesting on six thousand square meters of prime avian real estate. One-third of the Mediterranean greater flamingos will come to spend time in the Gediz Delta. Most are migratory, yet many others prefer to stay in the Gediz Delta year-round. The nesting island was constructed after local uni-versity scientists observed that the other flamingo breeding islands in the delta were rapidly undergoing erosion. The project was sponsored by the Gen-eral Directorate of Nature Conservation and National Parks (DKMP) and the municipal wetland management agency, IzKuş (the Conservation and Devel-opment Association of Izmir's Bird Paradise). The Izmir Municipality agreed to repurpose dredging boats to build the bird island, using the mud and sedi-ments extracted from the bottom of Izmir Bay.

By 2012, the island awaited the seasonal arrival of the birds. The then Democratic Party's mayor of Izmir, Aziz Kocaoğlu, described the flamingo island as one of his many infrastructural improvements for the city. At the inauguration ceremony in March 2012 Kocaoğlu proudly declared to the local press that "the city continues its arrangements to protect the Bird Paradise of İzmir . . . and to provide the most suitable conditions for its residents." Fea-turing 6,460 square meters of residential space for the birds, stretching 150 meters long, the flamingo island was "Turkey's first, and the third biggest in the world."[16]

In the spring of 2014, over seven thousand flamingos would hatch in the artificial island, attracting more media attention. The bird had become an icon of wetland management and, as later sections of this chapter show, of middle-class residents' moral attachment to the delta's livelihoods. But every-day decision-making about what the wetland was, what it should be, and what infrastructural transformation would craft a desirable ecology were embed-ded in an emergent network of contingent alliances, institutional collabora-tions, and power struggles, even as they also involved transnational negotia-tions over environmental meaning, science, and power.

Translations and Mosaics

Like the migratory birds, the Tour du Valat team came to the delta twice a year, following the organizational cycles of seasonal project funding. In the management headquarters, André, Bastien, and Charles (male French scientists in their early forties and fifties) were working intently on their laptops while I chatted with Lisa, who headed Tour du Valat's Gediz Delta project. Lisa was a lively and friendly American in her mid-forties, with red hair tied in a ponytail and sporting a loose cotton shirt and green hiking pants. Lisa talked to her Turkish collaborators in English and French, interspersed with some Turkish. She had arrived at Tour du Valat after a decade working in the humanitarian sector, which she described as a heavily taxing period of her life. Working at Tour du Valat, she told me, had allowed her the stability to raise a family.

In the Gediz Delta, Lisa's team collaborated with the environmental NGO Doğa Derneği, the wetland management agency IzKuş, and the University of Izmir. Her team's goal was to carry out scientific research in the wetland area and promote "integrated" and "adaptive" practices of wetland management.[17] Lisa described her French colleagues as "naturalists," obsessed with the natural world. But she was mostly fascinated with the relationships and "fields of interactions," as she put it, that constitute landscapes. Lisa had assembled her team over time, from people who were passionate and excited about working in the Gediz Delta—a former team member had been moved from the project because he was not committed enough, she told me. Enthusiasm and care were requisite for the job.

Lisa talked about the Gediz Delta wetland as a *mosaic* of different and coexisting habitats. Different "stakeholders," she explained, should be able to decide which habitats they prefer to favor and how the pieces of the mosaic (different patches of habitat) would fit together. The final shape of the mosaic would be a patchwork of social and political compromise—materialized in hydrological, ecological, and infrastructural environmental forms. Lisa was hoping to help form a Gediz Delta "platform" where all the different stakeholders could meet to discuss their visions for the wetland and share their research findings and priorities, eventually coming to a consensus on wetland management.

It was apparent from our many conversations during this and other visits that Lisa also was well aware of the complex power conflicts, personal grievances, and hierarchies at play between different institutions and individuals. The stakeholders sometimes refuse to collaborate, she told me on several occasions, or even to share their research data with one another. Further, the local environmental NGO she hoped would be able to form such a platform had remained quite marginal in actual decision-making about environmental governance in the delta, as were other civil-society organizations, including those representing the delta's villagers and fishers.

Tour du Valat had played a central role in the 1960s in stabilizing the very category "wetland" and in transforming it into a global object of environmental conservation, stewarding Turkey's efforts to set up a framework of wetland research and legal apparatus for conservation. More recently, beginning in 2007, I learned from Lisa and her team, Tour du Valat had been involved in a collaboration with Izmir's National Parks Department, the wetland management agency, the university, and several environmental NGOs. This was part of a wider "cooperative action" set up between Izmir Province and the PACA region (Provence-Alpes-Côte d'Azur) in France. Through this collaboration, Turkish and French scientists were able to frame the Izmir wetlands as an international object of environmental governance and, simultaneously, a model for wetland commensurability across Mediterranean sites, even as the Gediz Delta wetlands were simultaneously implicated in layers of national and municipal environmental governance and rife with conflict at all levels.

The Gediz Delta was also one of two field sites for Lisa's doctoral thesis. At the time of my fieldwork, Lisa was working on a social-scientific comparison of the intersection of different social networks of stakeholders in the Gediz Delta and in the French Camargue wetlands. One of the starting points of this work, as she explained to me on that rainy day in the wetland headquarters' cafeteria, was the realization that guidelines for integrated management for Mediterranean wetlands could not be equally applicable in all sites, because of, as she understood it, different institutional, political, and cultural structures.[18] One of the goals for this Tour du Valat's visit was to help IzKuş introduce more site-specific guidelines in the Gediz Delta management plan.

IzKuş, shorthand for the words "Izmir" and "Kuş" (bird), was a cooperative or association (birlik) formed between four municipalities—Foça, Menemen, Çiğli, and Izmir. IzKuş was charged with managing environmental conservation activities in the Gediz Delta's eight thousand hectares of protected lagoons, decommissioned saltpans, canals, lagoon mudflats, reeds, wet meadows, and small natural and artificial islands. In the delta's management headquarters, the scientific and technical staff of IzKuş shared office space with the DKMP's hunt-control division. Heading the DKMP staff in the conservation area was a friendly man in his mid-fifties. Before this assignment, he had been trained as an engineer and had worked as a manager in the state-owned saltworks at Çamaltı until the early 2000s, when the saltworks were privatized. He shared an office with Emre, the IzKuş biologist, in a spacious room on the second floor overlooking the salt marshes. Next to Emre's office was a forester working for the nature conservation district and a friendly accountant. There were no women among the technical and managerial staff. The IzKuş agency contracted a few delta village men to work as site guards and drivers and two village women to work as cleaning staff and sell tea and snacks in the small cafeteria.

The rain kept pouring, so Lisa, Charles, Emre, and I convened at a large table in Emre's office. Charles, who was a specialist in wetland management, showed Emre, on his laptop, a thirty-page document in French. This was the management plan for a section of Tour du Valat called Du Verdier, which, Charles said, presented habitats and kinds of land uses similar and comparable to those in the Gediz Delta's conservation area. Emre could write a similar document detailing the short- and medium-term goals for the Bird Paradise, he and Lisa suggested. To help with the conceptual translation of the management format from one delta to the other, Lisa and Charles asked me to translate the section titles of the Tour du Valat plan from French into English. Emre would then fill in this structure with the specific habitat, land use, conservation, and scientific goals for the Gediz Delta (which would be, in turn, translated into Turkish). Emre spent the rest of the afternoon intently working on the plan at his computer with Charles and Lisa, while I worked on translating an article on the delta's amphibian population from Turkish to English. However, Emre remained somewhat skeptical of the usefulness of this exercise. His visions for the delta did not matter, he told me on another day; the specifications he had added, working with Charles over two days, would never be implemented.

A Natural and National Wetland

The transformation of the Gediz Delta's agricultural fields, saltworks, wet meadows, lagoons, salt marshes, docks, reed beds, lakes, canals, and mudlands into a wetland began in the 1980s. Local ornithologists and hunters mobilized against the expansion of the Çamaltı Saltworks on nearby salt marshes. The saltworks, however, were not a recent development: people had been extracting salt in the region for centuries. In the eighteenth century, Izmir salt became an export good on global trade routes, shipped from the nearby port of Eski Foça on the northern side of Izmir Bay. Ottoman state authorities controlled extraction, which was auctioned to local tax farmers. Ottoman Greeks ran the mines and saltpans. In Eski Foça, salt from Çamaltı, Adatepe, and other sites was loaded on ships or transported on camelback to the rail station at Menemen.[19] In the late nineteenth century, the Çamaltı saltpans, previously allotted to individual workers or families, were turned into a centralized production system run by a private company.[20] After the founding of the Turkish republic in 1923, the Çamaltı Saltworks became a state monopoly. Salt was sold only on national markets, and production decreased.[21] As in the 1960s, Çamaltı salt today continues to be used in the petrochemical, textile, soda, chlorine, and leather industries.[22]

In the early 1980s, ornithologist Mehmet Sıkı conducted fieldwork research for his doctoral thesis in the Çamaltı Saltworks, in nearby Homa Lagoon, and in the surrounding marshes. Sıkı identified 183 bird species—50 of which were laying eggs in the delta's reed-bed habitats.[23] Sıkı and a colleague published a plea in the Turkish journal *Bilim ve Teknik* (Science and technology): "We expect that the Çamaltı management will recognize the protection of the wetland area inside the saltpans, which contributes with its unique and special value to the beauty and wealth of our country. We urge management to undertake this important duty and to stop the expansion of the saltworks." As a habitat for hundreds of species of birds, the scientists claimed, it was imperative to establish a protected wetland area of natural and national value.[24]

This advocacy work resulted in a series of overlapping conservation statuses for the delta's coasts. In 1982, Turkey's Directorate of National Parks and Hunting approved the creation of the Homa Lagoon Wildlife Conservation Area and Waterbird Protection and Reproduction Area. In 1985, the Ministry of Culture granted the same area protected status for its archeological remains and natural value. Izmir Municipality also planted eucalyptus trees to begin reforestation in the southern marshes. In 1987, the coastal delta was named Izmir's Bird Paradise.

In 1994, the Turkish government signed the Ramsar Convention for the Conservation of Wetlands of International Importance. Four years later, the

Gediz Delta became a Ramsar wetland together with three other Turkish wetlands, gaining new international visibility. In the early 2000s, Turkey's environmental legislation came to formally encompass wetlands. Wetland commissions, charged with environmental conservation and regulation of construction and resource extraction, were established in each province of Turkey. In Ankara, the ministerial Wetland Bureau began drafting site-specific management plans for all of the nation's wetlands.

In appealing to the "national value" of the delta rather than conjuring a planetary scale, the international networks of wetlands connected by the routes of birds, fish, insects, plants, and scientists, Sıkı situated the Gediz Delta wetlands as part of a Turkish nationalist discourse on nature's value—one that could be appraised primarily through scientific work. This formulation also reversed an older Turkish paradigm that posited green forests as an important symbol of national civilization, while unproductive marshes and deserts represented backwardness.[25] In fact, Sıkı would also campaign for the removal of a eucalyptus forest planted in the marshes during the rehabilitation efforts of the 1980s and for eradicating other nonnative trees and shrubs from the area.

Beyond appeals to nativism and nationalism, Sıkı narrated his role in the conservation of the wetlands in explicitly moral-ecological terms. The fate of the wetland exercised an ethical appeal on his own actions and on the wider purpose of his scientific work. For example, in 1994 Sıkı wrote on the expansion of the saltworks on the marshes: "I was deeply affected, and I believed that I could not remain an outside observer."[26] One morning in the winter of 2014, when I interviewed him in his university office, Sıkı conveyed the same sentiment of affliction for the fate of the marsh. He described being ethically summoned to intervene to reduce the destructive effects of increased drought, urban encroachment, and erosion of the delta's lagoons.

Sıkı's and others' appeals led to the creation of the multiple conservation statuses in practically overlapping areas. The porousness and multiplicity of these boundaries created a new space for regulatory contestation at the edge of the wetland. Residents, scientists, NGO workers, and bureaucrats who confronted fluid wetland boundaries strategically employed these hydrological and legal ambiguities to actuate their own moral-ecological visions. The legal performance of conservation statuses also engendered material transformations in the delta's wetlands, transformations that were shaped by scientists' moral visions of multispecies wetland ecologies.

Alongside university scientists such as Sıkı and elected officials and bureaucrats, Turkish environmental NGOs began to take an interest in the fate of the wetlands, advocating for their conservation by various means and filing lawsuits against new construction within the protected areas. These groups, however, also staked competing claims on the changing ecological infrastructures of the delta. Central to these contestations were evaluations of the quality and quantity of water in the wetlands, the rhythms of its flows, and the sediments it carried. These notions of water also conjured the broader political economy of the agro-industrial delta plain, inflected by the varying collaborations and tensions between the international wetland research network, university scientists, and the NGO staff.

Unmaking a Wetland

The day after I had helped Emre translate the Tour du Valat management model into a new template for his visions of the Gediz Delta's mosaic, in the late morning the French scientists, the NGO spokesperson, Emre, and I drove from the sports hall where the research team was lodged to the delta management headquarters. We arrived just as a group of men, all sporting business suits, were leaving the building's conference room. While rolling a cigarette, Deniz, the NGO spokesperson, pointed out some of the "main characters" as they stood on the front porch of the building, chatting informally while waiting for their staff to retrieve the cars from the parking lot nearby. Here were the director of the Çamaltı Saltworks, staff from the municipal water agency, the director of the National Parks Bureau with his entourage, and university ecologists.

Emre, Deniz, and Lisa had not been informed about or invited to attend the meeting. Deniz told Lisa that the meeting had been about planning a new bird-nesting island in Homa Lagoon near the saltpan. Lisa walked up the steps to the front porch, cornered the saltpan manager, and told him, in English, that she believed that before building a new artificial breeding island, there should be a comprehensive and holistic assessment on the impact on other birds' habitats in the delta. Rather than turn their attention to constructing a new artificial island, she urged, wetland managers should make sure the flamingo breeding island in the saltpans was effective in attracting a higher number of nesting flamingos. Then they should prioritize plans and budgets for ongoing infrastructural upkeep on the existing island rather than construct a new one.

Other contestations between communities of wetland advocates and experts centered on the materiality of the wetland's water, actuated through infrastructure work, as enabling ecological relations and livelihoods. One day in his field office, Emre discussed with one of the Tour du Valat scientists a new plan for introducing more freshwater into the Gediz Delta's wetlands— the plan had been espoused by the university's natural scientists. This was not the first vision of irrigating the Gediz Delta wetlands. Sıkı had been concerned about the drying of the marshes since the early 1990s, when he started attracting media attention over the effects of drought on marsh-bird habitats and petitioned government officials to introduce freshwater into the wetland.[27] His plan mobilized the existing infrastructure of agricultural irrigation—an infrastructure that had expanded to the delta beginning in the 1960s to support intensive agriculture of cash crops—to sustain the conservation area on the coastal wetlands. Similarly, the existing and planned bird-nesting islands leveraged the existing infrastructure of the saltworks and of the aquaculture infrastructure of a university-managed lagoon.

The Tour du Valat team generally agreed with Emre that pumping more freshwater into the wetland would not necessarily result in an improved habitat. During one of our conversations, Lisa emphasized that this approach risked damaging other habitats in the wetland, privileging freshwater marshes and the species of birds that thrived there over other pieces of the wetland mosaic. Charles added that the introduction of freshwater would destroy the typical Mediterranean ecology of the coastal marshes, which are environments characterized by salty water. And some habitats do not need to be covered in water year-round, he added, as the wetland ecology is characterized by cyclical and seasonal patterns of wet and dry (what hydrologists call a "water regime"). Charles suggested creating small "controlled areas" to evaluate the impact of freshwater introduction some years later.

The French team, environmental NGO, university, wetland management institution, and other groups of experts held contrasting understandings about what a wetland was, how it should be valued, and whom the wetland was for. At this level of everyday management and field science, the wetland appeared a much less stable boundary object than the one created through the international agreements and scientific collaborations of the twentieth century. Rather, the wetland was refracted through divergent moral notions of ecology. Contestations revolved around understandings of the meaning of the wetland as a mixed landscape, the kinds of nonhuman and human livelihoods that would thrive in it, and the material infrastructures that would support it. The contestations were not simply symbolic of different ecological ontologies, fields of expertise, and power relations between different groups of environmental managers and experts but had practical implications for the wetland that would be built and maintained.

The Reed Habitat Map and the Shifting Boundary of Freshwater

The Tour du Valat team came to the Gediz Delta twice a year and stayed for a week, but Emre, their closest local collaborator, worked in the wetland year-round. When I first moved to Izmir in 2013, Emre and his wife, a kindergarten teacher, lived in a small apartment in an urban neighborhood of Izmir. Emre was the only biologist at IzKuş. He had first started working for IzKuş as an environmental educator and tour guide ten years before, after graduating from college. At his university's birding club, where he met his wife, Emre had learned about bird-watching and developed what would become a lifelong passion for ornithology.

Emre occupied a rare position at the intersection of different institutions. When I first met him, he was completing a PhD thesis on storks nesting in the Gediz Delta, working at the University of Izmir with one of the most renowned conservation science experts of the delta. At the same time, Emre was a member of the environmental civil-society organization Doğa Derneği, the Turkish partner of Bird Life International. With his membership in the municipal association IzKuş, the university, and the environmental NGO and his long-term relationship with the Tour du Valat researchers, Emre was a well-liked figure. He mediated between people situated in each of these constituencies who often carried opposing visions of the wetland and contrasting goals. Over time, Emre also became an invaluable and knowledgeable interlocutor in my own research.

Every morning, after having a quick breakfast with his wife and feeding the house cat, Emre waited outside their tall apartment block in a sprawling urban neighborhood in Çiğli. He wore hiking pants and shoes complementing a wind jacket in green stripes and kept his long hair in a ponytail. In his backpack were a laptop computer, binoculars, a bird-watcher's guide, a pouch of tobacco and rolling paper, and a simple packed lunch—usually fresh bread or *simit* (a thin circular bread with sesame seeds), tomatoes, peppers, cheese, and olives. The driver, in a uniform of khaki shirt and pants, plodded relentlessly through the morning traffic, stopping for a few others on the way: the forest engineer, the accountant, and—sometimes—a foreign anthropologist. The car sped up on the motorway toward the Bird Paradise protected area in the Gediz Delta, and we parked in front of the wood and concrete building that served as management headquarters and visitors' center. Emre also enjoyed going to work on his mountain bike, skirting fast-moving vehicles on the clogged urban roads and ferocious packs of dogs on the roads alongside the delta's canals and industrial areas.

One day, Deniz and I went to visit Emre at his wetland office. He showed us his work in progress: a habitat map was overlaid on satellite imagery of the delta's coastal marshes. Walking in the marshes, GPS device in hand, Emre had worked for months to trace the edge of the reed beds, marking where freshwater gave way to higher concentration of salinity to complete the map. Deniz and I had accompanied him on several of these expeditions until a herd of feral horses had chased us away from the wet meadows. On the Google Earth map of the delta, the meandering line Emre traced with his steps in the deep mud transformed patches of reeds into a tangible mark of the ephemeral threshold of different kinds of water: water in the salt marshes and freshwater from the inland marshes mixing with irrigation water that was pumped into the conservation wetlands to keep them saturated and less salty.

Emre's meandering GIS trace included, as he pointed out to us on the computer screen, a long straight line showing where a wild boar had chased him away from the reeds. The three of us smiled at the juxtaposition. The geolocated cartographic representation of a human-boar encounter conjured the experience of wetland fieldwork as a situated and embodied narrative of place. But the cartographic representation of the boar encounter, intelligible to others in its GIS form only as Emre narrated it, would be cleaned up, erased,[28] and transformed into a document proving the ecological reality of changes in the quality of waters in the wetland over time. After all, Emre was tracking the moving frontier of reeds as a proxy for freshwater percolating on the salt marshes, not his multispecies encounters. Yet GIS habitat mapping, as a kind of "skilled vision,"[29] produced novel experiences of the delta that in turn informed the moral and political leverage of the representations produced by scientific exercises in skilled vision. Here, scientific practices sanctioned moral claims about the value of saltwater habitats over an irrigated freshwater wetland.

Emre's knowledge of the delta was based on his doctoral research work, but just as much on years of living and working in the delta, developing emotional and personal attachments to its human and nonhuman inhabitants: what historian of science Robert Kohler has called "residential" scientific practices.[30] One day Emre told me he had helped a delta village's headmen draft petitions to the National Parks Bureau to demand grazing rights for cows and sheep in the protected area (to no avail). Emre traced the delta's landscapes and social relations, which he knew intimately and in an embodied matter, into the more abstract representations of ecological transects, bird counts, maps, diagrams, scientific papers, and data sets. However, he moved swiftly between modes of engagement in the delta, connecting coordinates to stories, bird count data to poetic accounts of the everyday lives of birds, species lists to his practical knowledge of the delta's environments as they changed through the seasons. Like the delta's farmers, Emre knew what to forage; whenever I accompanied him during a fieldwork outing, he gathered wild mushrooms, asparagus, *Salicornia*, almonds, and bitter and tangy leaves.

Emre's habitat map, part of his collaboration with Tour du Valat, helped constitute the delta's wetlands as a site of "endangerment," drawing in my analysis here from anthropologist Timothy Choy's use of the concept to refer to affective commitment of scientific care for the preservation of threatened forms of life.[31] Emre's map provided a way of addressing more complicated questions about the historical ecology of water and irrigation in the conservation area—and the use of irrigation infrastructure to water the wetland. In deciding what to map, and where and when, Emre conjured his own lived knowledge and sense of place into cartography. Documents like the reed map, which made the Gediz Delta's habitats comparable with other wetlands, also framed the Gediz Delta as part of a transnational network of protected wetland ecologies. The Gediz Delta was not only Turkish but also a Mediterranean wetland and a wetland of "international importance."

When Wetlands Lose Their Functions

One day in winter of 2014, I rode the metro from my house in Alsancak to the university station. Adnan Kaplan, a landscape architect, greeted me outside the imposing Agriculture Faculty Building and walked me to his office on the first floor. On his desk Kaplan had prepared two small plates filled with sliced apples, pumpkin seeds, dried carrot paste, and hazelnuts. He had also set aside a tall pile of books, articles and reports on the Gediz Delta. Sipping several glasses of tea, we talked about the relationship between architecture and anthropology and about shared acquaintances at my university in the United States. Flicking through the reports and articles on the desks, we engaged in a five-hour conversation on wetland planning, interrupted occasionally by students from his architecture studio class, who eventually called him in for a review session.

I had come to talk with Kaplan because he had been leading research efforts on the "ecosystem services" of the Gediz Delta's wetlands.[32] He had also been a contributor to the 2007 Gediz Wetland management plan—a document realized on behalf of the Nature Conservation and National Parks Bureau in collaboration with two local environmental NGOs, Doğa Derneği and Ege Doğa.[33] During our conversation, Kaplan told me that he looked at the environmental problems in the Gediz Delta at a "landscape scale." The wetland, he explained, is a "mesh" of ecological structures and social relations. Kaplan was especially concerned about the pressure the northbound expansion of Izmir put on this mesh. This expansion was encroaching on the delta and damaging its wetlands.

"What happens when wetlands lose their functions?" Kaplan asked me rhetorically. He was implying that a wetland devoid of its labor—that is, of the work of providing ecosystem services—would no longer be a wetland. Drawing from the idiom of "ecosystem functions," a 1990s concept of "ecological economics" that seeks to find common metrics for disparate ecological processes, Kaplan mapped out wetland loss and degradation.[34] As part of his research, Kaplan had identified different landscape "types" in the Gediz Delta and attributed to each of them a ranking of functions (from "very high" to "very low") and anthropogenic stressors. Recently, he explained, he had also been working on issues of climate change, a topic that he found was still understudied in Turkey, lagging behind other countries where climate change had already become one of the primary concerns in wetland management.

I asked Kaplan what he meant by the term "wetland functions." His answer resonated with standard understandings of ecosystem function in the wetland conservation literature.[35] He listed a series of functions: for example, habitat, recharge of underground water flows, flood and climate control, food production, recreation, and aesthetics. Alongside ecological functions, the well-being of communities in wetland areas is crucial, Kaplan emphasized. I asked Kaplan what "community" meant in this context. He meant, quite simply, wetland residents and users.

Figure 2. New apartment buildings confound the boundary of the Gediz Delta wetland conservation area in Izmir, 2018. Photo by Benjamin Siegel.

I asked him if his team had integrated tools of community participation in making the management plan. I was thinking about all the disparate people and groups living and working near the wetlands who had now become "wetland communities" through the social-scientific language of the wetland management plan. In my questions, I was also trying to mimic the conventional language of participatory conservation, wondering whether it would resonate with Kaplan, but he shrugged and dismissed my question.[36] Instead, he mentioned a large survey study that his colleagues had conducted for the preparation of the management plan and stated that, of course, *all of* the rural residents' views had been taken into account, through the survey—as if to reassure me.[37]

As political scientist Begüm Adalet has argued, social surveys in Turkey have been key to the creation of modernization theory in the twentieth century, as they produced categories of modernity and traditions through survey respondents' answers and simultaneously sought to form modern subjects able to both administer and correctly respond to survey questions.[38] In this case, the survey had merely been a technocratic exercise. In a conversation with several men in the delta village of Sasalı I had in the summer of 2013, the elected village head and the president of the hunting association remembered the survey sheets that had been distributed and filled out cursorily by "household heads." They were surprised I had not come to deliver another survey. In our conversation, the two men and others had bemoaned that they had not been consulted in any meaningful way during the preparation of the management plan, and the wetland conservation project contained no provision that would benefit them in any capacity, other than for the few men and women hired to work in the headquarters.

Toward the end of our meeting, Kaplan recommended I read the work of the Turkish geographer Besim Darkot, which he considered foundational to understanding the Gediz Delta today. I was puzzled by the reference to a 1930s geographer but decided to follow his recommendation. The day after, I walked from my house in Alsancak along the Kordon promenade to the National Library in Konak to find Darkot's books and articles. I was very fond of the 1933 neoclassical building, with its yellow stone walls, arched windows, and green dome. Inside, I could always find an empty chair at one of the fifty wooden tables in the library's reading room, largely occupied by high schoolers studying for their university admission test and university graduates preparing civil-service exams. I enjoyed looking for my sources in the squeaky drawers of the card catalogues, a rapidly vanishing technology. It was easy to locate Darkot's oeuvre in the card cabinet, and the archivist promptly handed me a pile of his publications.

Politics of Sediment

In his 1938 *Coğrafi Araştırmalar* (Geographical research), a study of Turkish geography from the early Republican era, Besim Darkot described the hydrology of the Gediz Delta. "It is a known phenomenon that watercourses flowing in the alluvial bed change course frequently on their own, and the alluvial sediments brought by the floods explain the carving of the riverbed." Even as he wrote about past centuries, Darkot used the term "Gediz" to refer to the previous name of the river, Hermos, a name of Greek etymology.[39] The river name was changed as part of a wider ethnonationalist renaming of Turkish geography, which began in 1915 but was actuated in earnest after 1923. Topographical places and settlements, as well as languages belonging to non-Muslim communities, such as Armenian, Kurdish, Greek, or Bulgarian, were rendered Turkish.[40] Name changes went alongside the elision of cosmopolitan histories of anthropogenic deltaic environments, as scientists and planners recast them over the course of the late twentieth and early twenty-first centuries as "Turkish nature."

In Darkot's account, the Gediz River was active, creating land from flows of sediments accumulating on the coastal flats. Darkot characterized the river as a capricious being, one that changes course on its own volition. This dynamic world of moving water, sediment, and sand was largely devoid of human presence. Darkot described the "rapidity in which the alluvial plain has formed in a short period inside a bay, and the frequent changes of course, which have pushed back the river bed repeatedly toward where it used to flow before a former change of location."[41] Darkot's account conveyed the sense of a watery landscape in dynamic transformation, formed by the combined agencies of different kinds of water flows—in rivers, seas, and marshes. It also constructed a tale of agentive hydrological processes: one without history, people, and ecological relations.

Over the centuries, I learned, the Gediz River changed bed several times, moving increasingly southward toward the port city of Izmir. "If the shallowness of the bay explains the rapid land-making powers of the river, sediment makes it even shallower, threatening its function as a navigable city port, and thus calling for an infrastructural intervention to redirect the course of the river north of the city of Izmir," Darkot wrote. This is where Darkot reinserted human protagonists in his account, as skilled and ingenious engineers. In 1886, the Gediz River (then called Hermos) was redirected northward to prevent the delta's sedimented expansion into the bay. Without the redirection of the river, Darkot wrote, "the bay in a short period will become a lake, and the port of Izmir will disappear." The intervention saved the Izmir harbor from "the fatal destiny faced by other ports of the Aegean coasts."[42]

Darkot's story of the Gediz River's meandering closer and closer to Izmir Bay, until heroic engineers redirected the river to save Izmir, is a recurring tale I heard during the course of my fieldwork in my conversations with state officials, conservation officers, and environmental NGO staff. Local politicians frequently invoked the story of redirection of the Gediz River to justify infrastructural work in the bay as well as in the wetland conservation area.

In February 2013, Izmir mayor Aziz Kocaoğlu stated at a press conference that the city had saved one of the last remaining natural lagoons in Izmir Bay by undertaking construction work that had attracted environmentalists' and fishermen's critiques. "We usually understand that environmental conservation is predicated on a logic of 'nonintervention.' But if we hadn't intervened, the bay would be filled with sediment. It would become like the ancient civilization of Efes," the mayor declared.[43]

At a routine budgetary meeting of the Izmir Water Authority in November 2014, Kocaoğlu again invoked the 1886 Gediz River project to bolster his support for dredging the bottom of the bay to deepen the navigation canals. The intervention was framed as saving the bay from a fatal destiny—paralleling Darkot's narrative. "This is the project that will save Izmir," Kocaoğlu said. "Why?" he continued. "It is the project that will prevent it from becoming like Efes. Because sediment is filling Izmir Bay. All the data and the scientific evidence show that. According to a scientific estimate, a hundred years from now, the bay won't exist any longer."[44]

City officials and environmental managers implicated the Gediz Delta wetland in varying understandings of the changing infrastructural form of the city of Izmir. The delta, a constructed space of Turkish nature, was conjured materially, as built environment and ecological form, through connective infrastructures like bike lanes, the calculative potential of ecosystem services conversion, and the hydrological dynamics of threatening yet controllable sediments. These infrastructural transformations of mud, silt, marshes, and swamps are usually associated with projects of terraforming: the remaking of saturated places of shallow water into stable sections of dry land neatly separated from waterways.

In her work on the making of Kolkata, historian Debjani Bhattacharyya has called attention to how the shifting materiality of swamps remains latent in cities even as planners and engineers implemented legal frameworks and infrastructure for reorganizing mobile marsh landscape into firm land and property boundaries.[45] In the Gediz Delta, as in many other sites of contemporary wetland conservation, some terraforming infrastructures were subsequently repurposed to sustain the remaining marshes, driven by moral ecological visions, imaginaries, and practices of the wetland.

The wetland helps us understand the disconnective materialities and logics of infrastructure, broadly conceived as the built environment that allows movement (a highway, for example, a port, or an electricity cable) and that which moves (for example, a train, piped water, or information), as well as the organization of people and nonhuman beings implicated in the work of infrastructure or engulfed in its effects.[46] Since the 1990s, international agencies, conservation organizations, architects, and engineers have recast wetlands as "natural infrastructure"—in the sense that they perform work (such as absorbing pollution, providing flood control, or sustaining fisheries) that can sometimes be commensurate in monetary terms with the work of other built infrastructure. But wetlands are also shaped by and, in turn, shape other kinds of interconnected infrastructures. Wetland ecologies exist, thrive, and die shaped by the processes animating infrastructures as varied as irrigation canals, roads, commercial forests, power plants, factories, fishing lagoons, and bridges. Wetlands' seeping quality, the ambiguity of their boundaries, and their seasonal rhythms make them evocative and important places to stake political claims about the materiality of moral ecologies.

The Almost-Empty Bike Path

Kaplan had also been one of the planners and supporters of a bike route that connected a wealthy coastal neighborhood of Izmir with the delta, a twenty-five-kilometer ride I often took to get to the conservation area. I was always the only one on the path, save for the occasional spandex-clad cyclist or, during the weekends, organized bike tours for families, sponsored by large supermarket chains. The bike route was an infrastructural intervention that aimed to increase the flow of wetland visitors. These were envisioned by planners as able-bodied urban middle classes who would bike or walk to the wetland during their leisure time to practice nature observations, bird-watching, and sports.

The bike path was an infrastructure of "natural leisure," as Izmir city planners often described it. Connecting with the waterfront parks of Bostanlı's Mavişehir neighborhood (meaning, paradoxically, "blue city"), the path proceeded alongside shallow coastal marshes teeming with flamingos and the site of an eroded lagoon, dredged to its disappearance in the 1980s. Then it moved past a forested picnic area, a reforestation zone, and a zoological garden, arriving at the monumental entrance of the Bird Paradise, marked with a giant sculpture of flamingo eggs. It ended at the visitors' center, where there was also a hub for Izmir's new bikeshare system.

As I rode along the path, it was hard not to notice overlapping layers of infrastructure: the path left the controversial seaside villas in Mavişehir, which often flooded in the winter since they were constructed on drained wetland and landfill; a series of luxury high-rises; a shopping mall; two tall abandoned skyscrapers, which had been illegally constructed within the Ramsar boundaries, an industrial area developed in the 1990s; the city's wastewater plant; a former NATO base; a military airport; an airplane wing factory; and the Çamaltı Saltworks; and then it continued alongside a large drainage canal.

The bike path privileged a particular mode of middle-class leisure time, inviting visitors to the wetland at the same time that rural residents were excluded from grazing, foraging, and fishing within it. It was a visible sign of the more recent transformation of the Gediz wetlands from a zone of constructed wilderness and halted development to a space directly connected to new spaces of urban leisure—at the same time that wetland ecologies were threatened by ongoing urban, agricultural, and industrial expansion, including the construction of satellite residential developments. In a way, the bike route was promissory infrastructure. Bikes do not readily come to mind as a preferred means of wetland mobility: in the shallow waters, a human might move with flat barges, canoes, gondolas, or wading on foot. But the routes of boats are imperceptible after their passing: by contrast, bike paths, while disconnected from the wetland's saturated materialities and hydrophyte ecologies, form a visible and enduring mark.

In his ethnography of identity formation in post-Soviet Georgia, anthropologist Mathijs Pelkmans has written about "almost-empty" schools, shopping malls, harbors, and hospitals constructed during a neoliberal Georgian regime in the immediate post-Soviet period. Pelkmans argues that the almost emptiness of new buildings signaled the political dream of a transition to abundance and leisure, a transition occurring simultaneously with a neoliberal future and an imagined past.[47] Similarly, the almost-empty bike path embodied the uncomfortable space between a reality of inequality and chaotic redistribution of environmental pollution burdens and also an ecomodernist future of urban "greening" that is always just about to happen—a space that could be filled with different imaginations and aspirations of middle-class wetland mobilities and urban prestige.

Residual Feral Horses

Discussions between different groups of scientists over what constituted a desirable wetland mosaic and over the sediments of moral ecologies often centered on questions of nonhuman livelihood. These questions were implicated in the infrastructural rhythms of the rural and industrial delta, its networks of roads and canals, and the infrastructures of sediment control. Scientists mobilized specific ecological relations to advance moral arguments about particular infrastructural interventions in the wetlands.

A contestation over feral horses in the wetland exemplifies this point. Previously used by farmers for agricultural work and transportation, then abandoned in the wild, herds of feral horses (*yılkı atlar*) now graze and reproduce in the Gediz Delta, as in many other wetlands all over Turkey. Some advocates for the Gediz Delta consider horses to be relics of bygone agricultural systems, replaced by tractors, cars, and trucks. There was hardly any mention of feral horses in the management and scientific literature on the Kızılırmak Delta, and the voluminous Gediz Delta management plan mentioned horses only once.[48]

The advent of tractors and paved roads beginning in the 1960s had rendered horse work obsolete. These agro-economic shifts had created the condition of possibility for feral horses to become creatures out of place. With the expansion of irrigated agricultural land and new residential settlements, the delta's horse population was restricted to grazing in the mudlands, meadows, and hills of the wetland conservation area. One day, Deniz and I accompanied Emre on his reed-mapping fieldwork. A herd of about thirty feral horses appeared at a distance. We turned to observe them. For a few moments of multispecies intimacy, the horses stood still, and all looked at us. Then the horses approached, slowly at first, until they all began to run toward us in a cloud of dust, chasing us away until we reached the access road. This territorial behavior of horses, Emre told us, is a common feature of feral horse sociality.[49]

In 2011, the Tour du Valat research team conducted a field study on the delta's feral horses. They identified five separate herds with distinct social organizations and territorial behavior. In 2014, the researchers had estimated a population of around 150 horses and foals. However, local university biologists contended that the growing population of feral horses should be eradicated. At the least, their reproduction capacity should be controlled and curtailed. This anti-horse argument was based on supposed equine practices in the marsh: horses stomp on and destroy birds' nests and eggs and reduce certain birds' preferred nesting habitat of tall, thick grasses. Horses should make way for grassland birds.

In June 2011, Tour du Valat scientists, IzKuş staff, and Doğa Derneği's field biologists replicated an experiment that had been conducted at the Tour du Valat's estate in the French Camargue to evaluate the impact of horse grazing in the managed wetland. Scientists fenced an "exclosure" to prevent horses from grazing inside an area where plant species thrived in the seasonal flooding and salty soil: they identified *Juncus gerardii, Sarcocornia perennis, Spergularia* sp., *Scirpus maritimus, Aeluropus littoralis, Puccinellia festuciformis, Schoenoplectus litoralis,* and *Limonium* sp. The team monitored the exclosure during successive field visits, and in 2014, scientists conducted a final evaluation of the outcome of four years without horse grazing: the exclosure presented a reduction in plant diversity and had been completely overtaken by tall reeds.[50]

Domestic grazers—horses, cattle, sheep, goats—long held responsible for environmental degradation and encroachment into protected wilderness areas, had begun to be incorporated into environmental conservation projects worldwide in the late 1980s. Tour du Valat scientists endorsed grazing for its role in "creating and maintaining a mosaic of habitats." They also warned that "the effect of grazing is not univocal; it is complex and can produce benefits as well as negative consequences."[51] While grazing can favor the growth of certain plants, it hinders others. While it helps some species of birds, for instance, by creating open spaces in the thick reeds and reducing the impact of more aggressive species, it limits vegetation cover, thus reducing the habitat of other bird species.

The French team believed that local stakeholders should ultimately decide what kinds of habitat mosaic they want. Again, the Tour du Valat scientists connected habitat diversity to the democratic processes of decision-making. They were aware of the power hierarchies that privileged certain groups' moral ecologies of the wetlands. While this appealed to a postcolonial sensitivity to avoid marginalizing the work of their Turkish collaborators, Tour du Valat scientists nevertheless reproduced an existing hierarchy of scientific knowledge and bureaucratic power that shaped life and death in the wetland. The local farmers were meaningfully never consulted about their perspectives on and relationships with the feral horses.

The grazing experiment took on a life of its own and became embedded in contestations of wetland livelihood. In 2012, the director of the environmental NGO Doğa Derneği declared in an interview for a Turkish magazine that "the Gediz delta is a unique area protecting a variety of habitats, as well as people, birds, and horses." He explained that the "mosaic" habitat of the Gediz delta hosted 263 species of birds in the salty, freshwater, and brackish wetlands. The population of eighty feral horses, which farmers had abandoned in the wetlands, plays an important role in preserving wetland biodiversity. The NGO director backed his claims with scientific research showing that the grazing action of horses contributes to increased vegetation diversity. Preventing the horses from grazing, he warned, could result in 50 percent loss of plant species.[52]

Turkish conservation agencies rarely, if ever, integrated grazing into environmental conservation plans. With a few exceptions, cattle, sheep, and goats were supposed to remain outside forest and conservation areas. An agricultural engineering professor working in another Turkish wetland, the Kızılırmak Delta, explained to me that, in part, it was a problem of administrative coordination: even as recent management plans introduced provisions for establishing grazing protocols, the regional agricultural bureaus almost never followed up. He and others also often remarked on the lack of scientific research on the impact of grazing in wetland areas—research that, they hoped, could help dismiss earlier notions of the negative impacts of overgrazing. In fact, the prohibition to graze their animals in the conservation areas' meadows and forests continued to be a major concern for rural residents and had resulted in frequent acts of resistance against state officials and conservationists. One day, Emre showed me bullet holes in a sign marking the boundary of the Ramsar conservation area. Rural residents, he interpreted, had shot the signs as an act of protest against the enforcement of grazing and hunting restrictions.

Feral horses were a more complicated issue, which could not be framed as one of rural development or residents' claims over and access to resources in the wetland. Horses were at once wild and domestic—and this status could shift contingently.[53] Feral horses became domestic whenever conservation managers considered them to be an external nuisance: bureaucrats highlighted the fact that horses were farm animals and not wildlife. In the absence of current and traceable owners that state officials could hold accountable for the feral horses, however, the horses could become wild again. In this case, the responsibility of horse management in the conservation area fell on IzKuş and National Parks staff. Conservation officials did not have expertise to assess the horses' grazing impact. In fact, Tour du Valat's exclosure experiment had been the only one conducted in the delta. However, the relocation of large herds of horses, or their mass killing, would require complicated logistics and permits—for an issue that most people deemed largely inconsequential.

As a comparative example of animal removal, the control of urban stray dogs in Ottoman and Republican-era cities has long been a site of biopolitical interventions and violence. Techniques of segregation, deportation, death by starvation, and reproductive control were also used on human populations of unwanted subjects: religious and ethnic minorities and the urban poor.[54] In contrast to the mass killings and incarceration of Istanbul's and Cairo's dogs in late Ottoman periods and in the Turkish republic that historians and anthropologists have written about, however, the Gediz Delta's feral horses never gained the stable statuses of pests, outsiders, or impure subjects.

In early October 2013, the Gediz project team of Tour du Valat organized a scientific meeting at the IzKuş visitors' center, which I attended. The French research group would present their work in the delta and share their prelimi-nary results and recommendations. The issue of feral horses was going to be front and center at this meeting. Two days before the meeting, Lisa, Charles, Emre, and I convened at the large conference table on the upper floor of the wetland building. Charles reviewed with Emre the team's PowerPoint presen-tation for the upcoming meeting on the impact of grazing animals. Charles and Lisa explained to Emre and me the grazing policy in the wetland man-agement plan of the Tour du Valat estate: local farmers could apply and pay a modest monthly fee to introduce a limited number of cattle into the wetland, which grazed alongside the estate farm's horses. They insisted that Emre try to introduce a controlled grazing provision in the next iteration of the manage-ment plan. In a later conversation, Deniz, the spokesperson for Doğa Derneği in Izmir, told me that the presentation was a strategy to leverage foreign experts and the comparative example of the Camargue with powerful local proponents of the no-grazing approach. The issue was particularly tricky, since one of the opponents of the presence of feral horses was also Emre's own PhD supervisor.

The meeting was held on October 3, 2013. Before the conference, the Tour du Valat team had convened to discuss a strategy to bypass the power dynam-ics and disagreements between the wetland management, the university scientist, and the National Parks office. Lisa and Deniz agreed that the eldest member of the team, Bastien, would lead the potentially controversial presentation, comparing horse-grazing impact assessments in the Camargue and in the Gediz Delta. Deniz would translate his presentation from English into Turkish. I took a seat in the back row of the small auditorium. Among the attendees was the director of the Provincial Section of the National Parks and Nature Protection in Izmir with his office staff. The director was particularly interested in the horse-grazing research, since it directly concerned his mandate to control animal populations in the conservation area. He directed several follow-up questions to Bastien, who presented Tour du Valat's research and conclusion.

After the meeting, the controversy over the horses was picked up by local news reporters. In an article for the regional newspaper *Hürriyet Ege*, the National Parks director asserted that his department would address the issue of the horse population, which had clearly grown in recent years. He conceded that he had recently learned that the horses were not in fact damaging the marsh ecosystem. In the newspaper interview, the director talked about the horses as essential to wetland habitats and emphasized that the Bird Paradise Park should not just be seen merely as a bird habitat. "The Gediz Delta wetland is rich in biodiversity, and that includes wild animals. It also includes horses and meadows—they are also a component of the ecosystem," the director said. The problem, he continued, lay not with the naturally growing population of horses but with residents of nearby villages who continue to abandon their horses in the wetland conservation area, "despite all the measures that have been taken." The director concluded that his department "would conduct a study to determine how many horses will stay and how many will need to be expelled from the Bird Paradise" and install fences to prevent them from returning. Interviewed for the same article, Sıkı, the natural sciences professor, declared that the delta horses were "reproducing without any form of control," causing "great damage to the wetland ecosystem." He would not allow "turn[ing] the Bird Paradise over to horses," and he would "fight until the horses are removed from the area," applying to the public prosecutor's office, if necessary.[55]

On the one hand, convinced by the Tour du Valat researchers and by the comparative example of horses grazing in the Camargue wetland, the National Parks director had conceded that horses had become an essential part of the wetland ecosystem. The horses' reproduction, which the department had previously seen as a threat to wetland ecology, had then become a natural element of the wetland. On the other hand, the director leveraged the anthropogenic threat of the unruly villagers abandoning their worthless horses in the wetland to justify the removal of the "excess" population of horses and to construct a border fence to keep them from returning. This plan, however, never specified whether the horses would be killed or relocated, how they would be captured and transported, or how large the "ideal" population of horses was.

This is not a story about how scientists transform communities of animals into natural species for thriving biodiversity,[56] nor is it about the leveraging of charismatic megafauna to rouse environmental conservationists.[57] This is a story about the mundane politics of wetland jurisdiction, one naturalized through appeals to ecological processes. The stakes in these processes are material transformations in the wetlands: grasses that horses stop on and eat, destroyed nests, habitats more or less favorable for different species of birds. The horses then become an active infrastructural element in the wetland, although residual to previous agricultural infrastructures of work.

The absence of traceable legal owners among rural residents—responsible parties, that is, who could be fined and asked to remove the horses from the conservation boundaries—shifted the burden (and cost) of horse removal onto state and municipal authorities. IzKuş and National Parks staff routinely imposed hefty fines on the "illegal" hunters and cattle herders in the conservation area. The inclusion of horses, but not cows, sheep, or buffaloes, as legitimate wetland residents was contingent on these agro-economic dynamics. National Parks staff were able to appropriate the scientific argument that Tour du Valat scientists advanced over the neutral effects of horse grazing on bird habitat, in order to respond to the pressure university scientists had long exercised on the department to remove the horse population.

The adjudication over the horses' right to live in the wetland was largely symbolic of the sedimentation of meanings of nature in the wetland, which in the process was cast again and again as a wild, natural habitat—albeit one that had to be made and maintained through human work. It also performed the existence of a community of "wetland stakeholders," connected to prestigious international projects of wetland management while ultimately legitimizing the purview of the National Parks office over other visions. Anthropologists Deborah Gewertz and Fred Errington have advanced a similar argument about the bureaucratic procedure of endangered species protection in the context of river management in the United

States, a mechanism that results in its self-legitimization rather than prevents the death by extinction of an endangered fish.[58] By contrast, the delta's horses did not hold a protected status as an endangered species, nor did other wetland infrastructures threaten their livelihood. The horses' reproductive capacity constituted a threat to the putatively natural wetland ecosystem. As wetland managers cast horses as part of the wetland ecosystem, following international understandings of wetlands as habitats that were also maintained through grazing, the National Parks shifted focus on the rural villagers' "illegal" practices to render the horses, once again, a surplus population. Having reinstated his authority over the French researchers, even as he incorporated their insights into his own argumentation, the National Parks director's job was done. The feral horses, ever elusive, continued to graze, undisturbed, in the conservation area.

Conclusion

The materiality of the wetland—and especially, the movements of sediments and water—is ambiguous, contested, and always in the process of becoming something else. As scientists, NGO staff, and bureaucrats wrestle with what wetlands should be like, they make claims of desirable and moral ecologies grounded, literally, in the materiality of wetlands. This materiality is entangled with infrastructural form and function: wetlands seep through infrastructures as varied as irrigated fields, military airports, seawalls, mines, roads, and artificial islands. Arguments and imaginaries of moral ecology conjure the physical remaking of built environments. These remakings are sedimented in landscapes, and past layers of transformation interact with new ones in different ways.

Rather than use metaphors of environmental winds,[59] networks, or flows to describe the international production of the wetland through the work of governments, NGOs, scientists, and activists, I focus on sediments. Sediments are particles floating in liquid and depositing at the bottom. Sediments challenge the dichotomies of water and land that anthropologists and historians have so often focused on. In debating the moral meaning of the wetland, and the kinds of materialities that should form wetlands, scientists, bureaucrats, and NGO workers in the Gediz Delta wrestled with varied processes of sedimentation, of shaping wetland elements into different kinds of forms for different kinds of ecological world making.

A second thread running through this chapter concerns the disparate and contingent collaborations at different levels that produce the wetland as a natural, legal, political, and cultural object—a process of wetland making that is contested at each step. I followed this process at work during everyday practices of ecological fieldwork, conference presentations, press conferences, informal conversations, and seemingly inconsequential debates. This is the everyday life of wetland making, and for scientists, bureaucrats, and advocates the wetland is at once a proxy for local and national politics and a moral ecology that generates meaning of its own through the varied livelihoods it enables or restricts.

Scientists and elite hunters in Izmir advocated for the conservation of the Gediz Delta's coasts in response to the transformation of the remaining salt marshes into saltpans in the 1980s. During the time of my fieldwork, wetland managers debated how much freshwater should be pumped into the marshes to, literally, irrigate the wetland. They conjured different kinds of water flows of different qualities, enabled by already-existing infrastructures, and carrying different kinds of sediments. Specific wetlands ecologies resulting from this material and moral crafting of landscape would provide livable natures for different kinds of critters.

These politics of endangerment were not hegemonic. State officials, scientists, and advocates disagreed on who should inhabit the wetland. Rather than demonstrate a tension between contrasting kinds of care, care of the individual and care of the species, like those Thom van Dooren has written about,[60] here I emphasize that these practices of caring for the wetland are about the inextricable entanglements formed between *communities* of animals and built ecologies. By constructing new flamingo-nesting islands using sediments dredged from the bottom of the bay, city officials crafted new infrastructural ecologies. The construction of the flamingo nests also recalled the symbolic salience of construction sites in contemporary Turkey as a performance of political power. These nests reminded me that environmental conservation is always also an infrastructural project. At the same time, however, these nests also placed flamingos front and center as rightful wetland residents whose reproductive capacity in the newly constructed environment attests to a successful wetland conservation politics. In contrast, horses were cast as unwanted residues of a rural past, and their livelihood actions of walking and grazing in the marshes and grassland as a form of infrastructural labor that called for violent responses.

The Gediz Delta wetland was conjured and formed alongside other kinds of infrastructure—through connective bike lanes, ecosystem services calculations, and the controlled movement of potentially threatening sediments, such as the sediments that might clog waterways and close Izmir Bay from the sea, rendering it a shallow lagoon. Moral ecological claims are connected to the imagination and control of material processes. Mud, sediments, and seepage are political objects that endured imperial, colonial, and ethnonationalist projects of separating land and water. Projects of wetland building, however, remain ambiguous and contested, as I show in describing everyday debates between city officials, NGO scientists, university experts, and environmentalists. These contestations and ambiguities, rooted in diverging moral imperatives of wetland livelihood, unfolded over mundane infrastructural practices, such as pumping irrigation water into the wetland, eradicating nonnative trees, building a bike lane, fencing feral horses, and dredging mud from the bottom of the bay to build birds' nests.

3   Moral Ecologies of Infrastructure

The Delta Is Dead

When I first met Captain Barbaros, I imagined that I would sail with him from the fishermen's port in Izmir's Mavişehir neighborhood, where we would see the Aegean Sea merge with the salt marshes of the Gediz Delta. Instead, Barbaros asked me to join him on board a crowded municipal bus. As the vehicle climbed the steep hills, Barbaros pointed down at the luxury apartments near the docks where his boat was anchored. Before the middle-class complexes were built, he recalled, there had been small houses with gardens boasting delicious melons, all framed by the waterways, canals, and marshes where Barbaros fished for eels, gathered succulent *Salicornia* (*deniz börülcesi*), and shared his deltaic livelihood with hundreds of species of waterbirds. But now, he declared, after a long silence, "the Gediz Delta is dead." Barbaros spoke to me of the marsh drainage projects that the city of Izmir had undertaken in the 1970s. By the 1990s, the coastal marshes and gardens were gone, replaced by high-rises and luxury villas guarded by private security. Barbaros and his family had moved, too, to a modest apartment up the hill.[1]

Hoping to understand the Gediz Delta from Barbaros's boat, I had envisioned writing about a delta animated by the movement of fish, technology, sediments, boats, birds, plants, capital, bacteria, currents, and multiple waters, inspired by recent work on the anthropology of water.[2] This delta, like other watery environments in Turkey and beyond, had also been shaped by nineteenth- and twentieth-century state-led marsh reclamation projects aimed at obtaining new land for agriculture, industry, and cities; creating new populations of national subjects; and displacing nomadic marsh dwellers.[3] But in proclaiming the "death of the delta" from his hillside concrete apartment block, Captain Barbaros emphasized why, and to whom, these changes mattered. He assessed, in specifically moral terms, the new, precarious, and potentially mutually annihilating relations between communities of people, nonhuman plants, and animals, all enabled by changes in the delta's built environments. Barbaros, like my other interlocutors in Izmir, articulated a moral stance on the deltaic worlds of infrastructures and ecologies, one that posited them as at once entangled and inseparable.

This chapter discusses what is at stake in theorizing infrastructure and ecology as inseparable rather than set in opposition. Delta fishermen made visible their moral claims about the unjust effects of infrastructural transformation on their livelihoods and on delta ecologies, writing petitions, for example, and filing lawsuits. Distinctions between infrastructure and environment no longer hold; rather, they fold onto one another in complex interplays, writes anthropologist Kregg Hetherington.[4] For Barbaros, however, the delta did not die because it was suddenly made into infrastructure: the delta has long been infrastructural, for as long as human communities and their nonhuman companions have lived in the region. Nor did Barbaros find hope in critters thriving in the infrastructural rubble or in weeds growing in the cracks. What does matter—with a moral urgency—for him and for others are the contingent and unequal outcomes of particular infrastructural arrangements of organisms, materials, and economies. The moral ecologies of infrastructure discussed in this chapter also challenge binaries of urban and rural, resistance and hegemony, even wet and dry.

Expanding the concept of moral ecology beyond its previous meaning as, broadly, resistance to capitalist processes and dispossession, helps us move beyond a binary where ecological relations are posited as opposed to infrastructure—or, similarly, thriving in its interstices, disrupting engineers' plans and capitalist trajectories.[5] Moral understandings of ecology in the delta are not necessarily emancipatory, inclusive, progressive, or sustainable. They are nested within capitalist processes, histories of land expropriation, class and gender hierarchies, and exclusionary ethnonationalism.[6]

The analytics of moral ecology illuminate the urgency through which Turkish residents have tackled questions of environmental and infrastructural change. Understandings of ecology in contemporary Turkey are mediated by the politics of large-scale infrastructure development and by the rapid trans-formations of agrarian, mountain, coastal, palustrine, and urban environ-ments under the AKP (Justice and Development Party) government since it came to power in 2002.[7] These have variously been projects of regionalism and of national development.[8] Concerns over infrastructure (especially urban, water, energy, and transportation) have come to stand in for broader claims about livelihood and democracy and have also become vehicles for contesting state power.[9]

Here, however, I move away from a focus on Turkey's large (and small) environmental mobilizations, such as the June 2013 protests against the destruction of Gezi Park in Istanbul, as well as popular opposition to ther-mal and nuclear power plants, run-of-the-river hydropower, and mining.[10] I turn instead to everyday experiences and conflicts over infrastructural ecol-ogy in the making and unmaking of the Gediz Delta as a wetland, a fishery, an open-air laboratory, a transportation node, and a site for speculative real estate. While many of these changes are conditioned by national and regional policies, I foreground fishermen, scientists, NGO workers, and other work-ing- and middle-class residents in Izmir as they reimagine their own ecologi-cal entanglements. They advance varied moral claims on the livelihoods and relationships of humans and nonhumans that emerge from transformations in these sites and elsewhere in the delta. These moral evaluations of specific deltaic configurations speak to actors' participation in environmental pro-cesses. Their moral ecological articulations take various forms—for example, intimate conversations, lawsuits, scientific arguments, political declarations, and mapmaking. In turn, they enable further transformations of the delta's built environment and its wetland ecologies.

Immoral Infrastructures

Moral ecologies reflect varying understandings about who gets included and excluded in environmental decision-making and who reaps the benefits of infrastructural and ecological transformations. One morning in March 2014, two elderly fishermen and I drove past a new residential gated community in the lower Gediz Delta at the edge of a wetland conservation area. One of the fishermen, Ahmet, gestured to the cotton fields that ended at the edge of road leading to the half-built villas and to the ponds nearby suffocating in green algae. "This is Sıkı's doing," he uttered, referring to the university ornithologist who had campaigned to create a protected area in the delta in the 1980s. Yet Ahmet was not concerned with the marshes drying as water was redirected to agriculture or the loss of waterbird habitats to the expanding salt industry. Rather, Ahmet bemoaned the effects of conservation zoning and the elusive benefits of speculative urban growth. "The city of Izmir could have expanded here," he elaborated. "Then, there would have been plenty of jobs for us. But conservationists prohibited development. Now, there are only these small gated residences for the middle class, and they stand half-empty."

Ahmet's moral claim, concerning economy as much as ecology, resonated with those of other rural residents as they considered the construction of high-end gated communities since the 1990s. This urban, middle-class moral ecology counterpoised the "unhealthy" city with the "clean" countryside. Yet rural residents advanced a reverse moral ecology as they complained that the middle classes, in turn, were polluting their countryside.[11] But Ahmet was also concerned with the ecological effects of toxic industrial runoff flowing into the delta's coastal marshes where he fished.[12] Working-class residents like Ahmet and the urban middle class both employed shifting dichotomies of healthy/toxic and legal/illegal as indicators of just/unjust landscape making.[13] The making of these contrasts suffused divergent moral ecologies of delta infrastructure.

The other fisherman, Bekir, also found fault with the residential development. "Look," he declared to the two of us. "All these houses were built illegally. The building cooperative made a huge profit, but we won't. They should not be here. This is not environmental conservation! The delta's management plan was made by people who never even come here. They just looked at their computers in their air-conditioned offices. They never talked to us. They don't know the delta." Bekir articulated a discomfort with classed forms of environmental expertise: even as an experienced fisherman and the elected fishermen's cooperative representative, he could not participate in an equal dialogue with scientists and bureaucrats about the possible futures of the delta. Bekir was involved in a long struggle with the university and the conservation area's management over a nearby fishing lagoon, which I detail later. Contestations over infrastructure constitute new political collectivities, as anthropologist Leo Coleman has written.[14] People create and maintain social infrastructures—affective and political networks—to sustain social relations or to exclude others from forms of collective belonging.[15]

Conservation scientists, too, understood wetland ecologies as infrastructural, political, and moral. During an interview I conducted in his university office in winter 2014, Sıkı recounted "irrigating" the protected wetland following a moral imperative to safeguard Turkey's natural and national wealth and the livelihoods of hundreds of thousands of birds in the delta. Following his recommendation during periods of drought in the 1990s, the DSI built a water canal and installed a water pump to provide "much-needed" freshwater to the marshes. Sıkı had envisioned it as an underground conduit, but the DSI had instead constructed an open canal from which, Sıkı alleged, local villagers illegally (and immorally) extracted water upstream of the drying and dying marshes.

As pumps and canals irrigated the wetland, the lower delta's water was being redirected to sites of water-intensive industrial and agricultural production, which, in turn, released agricultural and industrial pollutants. The delta was thus simultaneously produced, sustained, and endangered through moving flows of water and sediments. Sıkı's vision of a wetland's moral ecology was predicated on infrastructural remakings privileging putatively native species. Irrigation infrastructure that contributed to drying the marshes, he believed, could also be used to the opposite effect. He saw the gated community under construction as constituting an obstacle to wetland conservation, even as the conservation area itself increased its real estate value.

These infrastructural ecologies presented different moral quandaries for residents, as my conversation with Ahmet and Bekir exemplified. Neither conceived of delta ecology outside its built infrastructure, yet they saw different connections and attributed contrasting moral salience to the changing delta. For Ahmet, the gated community and the toxic sludge flowing from upstream factories into the delta's fishing grounds were indicators that planners and managers had excluded people like him from their visions of the delta's future. For Bekir, the residences were a blatant example of the shortcomings of environmental conservation. They also epitomized the reckless practices of the Turkish construction industry. Both men worried that residential infrastructure in the delta left less space for their livelihoods and for their ability to make claims heard across bureaucratic and economic hierarchies. These were moral claims about the place of the rural working class in transformations of urban form, environmental governance, and agricultural infrastructure.

Moral Ecologies beyond Resistance

My understanding of moral ecology goes beyond dichotomies of anticapitalist resistance against "immoral" ecologies. Bekir and Ahmet did not articulate a critique of capitalist and neoliberal systems of resource use.[16] Rather, the fishermen remained embedded within them.[17] Neither did they posit infrastructure as destructive of ecology: in different ways, they understood them as coproduced. Anthropologists have often foregrounded the damaging effects of infrastructures—including river dams, orchard fences, landmines, and flood shelters—threatening the livelihood and existence of human communities, animals, plants, and agrarian landscapes.[18] Alternatively, studies have called attention to emergent ecologies that (somewhat unexpectedly) thrive in infrastructural rubble and ruins,[19] as well as in infrastructures that unintentionally become habitats for nonhuman creatures.[20] In these studies, ecological relations are often described as either vulnerable to or able to transcend infrastructure; conversely, infrastructure is generally cast as dominant, hegemonic, and oppressive.[21] By contrast, I see the *mutual constitution* of ecologies and infrastructures, rather than their opposition, as conduits for moral claims about human and nonhuman livelihood.[22] Many environmental historians have written about environments as mutually constituted with human-built infrastructures rather than apart from them.[23] The analytics of moral ecology helps account for the different ethical visions that suffuse ecological relations in their infrastructural transformations.[24]

Scholars have often cast moral *economies* as appeals for mutual responsibility and collective justice: calls for equitable access to land, for example, or fair remuneration for work.[25] In his analysis of eighteenth-century bread riots in England, historian E. P. Thompson argued that they expressed a collective consensus about legitimate and illegitimate practices of bread making and distribution, as well as communal norms and obligations.[26] Writing on peasant rebellions, political scientist James Scott redefined moral economy as a notion of economic justice that responded to the exploitation wrought by colonial transformations of labor and land.[27] In its extensive contemporary anthropological use, the concept of moral economy highlights the recognition and fostering of reciprocal economic obligations between social groups. For example, anthropologist Kimberly Hart has analyzed Turkish villagers' religious evaluations of the new capital and profit brought by their carpet-weaving activities as one particular manifestation of local moral economy.[28]

Moral economies are often seen as responding to "immoral" practices of profiteering and the alienation of capitalist markets and neoliberal restructurings. However, Thompson emphasized moral economies as a shared consensus about what constitutes legitimate or illegitimate practices: "definite, and passionately held, notions of the common weal."[29] Indeed, many anthropologists have emphasized the embeddedness of notions of morality to hegemonic political and economic transformations.[30] For example, Andrea Muehlebach's ethnography of Italian neoliberal transformations has foregrounded the ongoing relevance of moral practices to post-welfare economies, "a form of ethical living that appears as the negation of and yet is integral to neoliberalism more broadly conceived."[31] Recasting understandings of ethical subjecthood beyond its concern with exchange practices and economic relations, I theorize moral ecology to highlight mutual obligations, affective relations, and valuations among humans as well as plants, animals, and other organisms. It is an expanded moral economy that takes into account relations with other living beings, material transformations, and relations of symbiosis.

In recent anthropological scholarship, moral ecology has denoted practices protecting collective resources and sustainable and reciprocal relations between environments and societies and resisting capitalist and corporate expropriation.[32] This use, however, tends to reproduce a dialectic of ecological morality defined as resistance to the immorality of markets, states, or corporations. This ultimately results in a predetermined understanding of moral claims and fails to account for the situated norms and values embedded in capitalist, corporate, and neoliberal transformations of environmental relations. In my use of the term, moral ecologies do not simply denote traditional subsistence practices and indigenous ontologies counterposed with, for instance, high-yield seed varieties or bounded conservation zones.[33] Rather, I see these "modern" projects also as forms of moral practice and productive of new environmental subjectivities.[34] Fishermen's, city planners', and scientists' moral notions and claims emerge alongside (and do not simply resist) infrastructural transformations of the Gediz Delta and their active participation in these changes.

A Chronicle of Destruction

On a sunny spring morning, I was driving on a gravel road with my friend Emre, the biologist who worked for the wetland management institution, following a narrow strip of land separating a lagoon from the Aegean Sea. Delta residents knew the lagoon as "Homa," but a sign marked it as the "University Aquaculture Department's Lagoon." As we drove, Emre pointed at floating foam and trash, coming from new road construction, which had further separated sea and lagoon waters and drastically reduced circulation within the lagoon. An underground conduit connected the lagoon to an abandoned saltpan, where white and brown organic matter was pushed by the wind from the lagoon into the channel and accumulated in one corner. Our road ended at the fish traps: a long metal structure across the shallow mouth of the lagoon. Near the traps was a dilapidated one-story building.

We boarded a small inflatable boat to reach the other end of the sandbar, beyond the fish traps. Wading ashore, I struggled to keep up with Emre, our bare feet pricked by sharp shells and stones. Sitting down on a bush of red and green *Salicornia*, I made an inventory of trash: plastic flip-flops, Lycra and woolen rags, beer and water bottles. Meanwhile, where lagoon and sea waters merged, Emre was counting a colony of *deniz kırlangıcı* (Sandwich terns). Emre's perception was synesthetic: he used eyes, binoculars, ears, notebooks, and the intimate knowledge of this lagoon, earned through decades of fieldwork. "To recognize birds," he told me another day, "is to know birds the way you know a close relative, a friend, a loved one. You have to know their *ruh* [soul]."

Only sailing back at sunset did I realize that our work also chronicled a moral ecology of changing infrastructure. Homa's waters were rapidly becoming sea, as waves and wind eroded the lagoon sandbars. Emre explained that agricultural infrastructure had severed the lagoon from the Gediz River estuary. Removing dikes and drainage canals would once again open the delta's marshes to their seasonal hydrological pulsing and allow freshwater to flow into the lagoon. This, he acknowledged, would be impossible. But many species of birds continued to make the lagoon their home, and some infrastructural changes risked destroying their habitats, while others would allow for their flourishing. For example, the artificial islands built in the saltpans provided a popular nesting ground for thousands of flamingos, as mentioned earlier. However, environmental advocacy groups and delta fishermen heatedly contested different projects of lagoon restoration.

Our fieldwork would help Emre and friends at an environmental NGO draw a map showing bird-nesting sites that would be destroyed by the university management's restoration project. This was a moral vision of nonhuman livelihood (at the scale of flocks, ecological habitats, and species), entangled within a spectrum of more or less desirable infrastructural configurations, all of which were lived ecologies. Like the storks he had studied for his PhD thesis, Emre's home was in the liminal infrastructural ecologies at the edge of the wetland. He and his wife had recently moved from Izmir to a new middle-class residential community in the lower delta. Most other units lay empty, purchased as investments during Turkey's recent economic bubble, now burst. In this speculative infrastructure, they grew grapes, raised chickens, and fed the feral dogs. Alongside Emre's concern for the livelihoods of birds and fish, fishermen had been reclaiming different ecological infrastructures for the lagoon against the university-led creation of an open-air wetland laboratory.

The Lagoon Trap

When I first met him in 2014, Bekir, one of the two elderly fishermen and head of the delta villages' fishing cooperative, was leading a prolonged crusade against university managers who had prevented fishermen from accessing the Homa Lagoon fisheries. Over centuries, Homa had been made and remade around the seasonal flows of water and fish. Fishermen collectively maintained the lagoon using *kargı* (giant reed, *Arundo donax*), *saz* (phragmites, *Phragmites australis* and *Juncus* sp.), and stones. These were mobile infrastructures, which the fishermen moved seasonally. In early spring, schools of fish swim through a narrow inlet into the lagoon to feed and reproduce in the shallower, warmer, and less salty waters. In late fall, as lagoon waters cool, those fish prepare to swim out toward warmer and deeper seawater and get caught. Lagoons, in a sense, are traps.

One morning, over cups of tea and *börek* (a savory pastry) in his neighborhood's teahouse, Bekir told me that his parents had not been fishers; rather, they had always associated fishing with the Greek Orthodox villagers in the delta. From the work of historian Emre Erol, I later learned that many Greek Orthodox fishermen in the region were landless peasants who worked seasonally as sharecroppers, fishermen, and salt miners.[35] The name of the delta's biggest fishing lagoon, Bekir continued, was a Greek word: "Homa" meant "earth," a translation derived from its shallow and turbid quality. Greek Ottomans had in fact dominated maritime professions, such as sailing, fishing, and oaring, until the 1920s.[36] But unsurprisingly, Bekir did not mention the violent deportations and forced exchanges that led to the eradication of the Greek Orthodox population between 1914 and 1923.

In the late seventeenth century, Smyrna (the Greek name of Izmir) grew from a small port into a cosmopolitan city and global maritime trade node. Its surrounding landscapes produced export goods such as dry fruits, salt, and stone.[37] In the early twentieth century, the forced deportations and genocide of Greeks and Armenians, alongside war and occupation and population exchanges, dramatically transformed the demographics of Izmir.[38] Anthropologist Leyla Neyzi has argued that residents' memories of the violent remaking of Izmir from a cosmopolitan port city to a Turkish town emphasized narratives of loss even as they embraced wider nationalist stances.[39] Yet no scientist, activist, or fisherman I talked to had ever mentioned to me the Greek history of Homa—except Bekir, during that one breakfast conversation. In the remaking of the delta as a site of national value, older cosmopolitan histories of landscape use, particularly those uncomfortable to narratives of ethnic nation-state formation, had been elided, thereby showing that ethnonationalism is bound to moral claims about ecological and infrastructural form.

Bekir was born in the delta village of Tuzçullu, then known for sheep farming and camel caravans—carrying, among other goods, salt from the nearby mines. The family claimed nomadic *yörük* heritage, and in his childhood Bekir had been a shepherd. As a teenager, he told me, he had moved to the expanding industrial outskirts of Izmir to work in a paper factory. After retirement, he bought a small fishing boat, nets, and other equipment and joined the local fishermen's cooperative. Bekir's son, too, had became a full-time fisherman only after quitting his job in an auto shop in the wake of a fight with his boss. Only at sea, the son told me, lighting a cigarette as I pulled in nets of squid, algae, and crustaceans to the boat, did he feel free and at peace. From the deck, he threw small fish into pelicans' open beaks. The birds knew to follow his boat, he said, and would be ready to jump and swallow the fish.

Bekir's family story resembles those of many other landless (or smallholder) peasants in the lower delta: old and young men (and a few women) at sea during the fishing season, working flexible or seasonal jobs and, if they have the capital, opening fish restaurants in new lower-middle-class compounds. Tuzçullu fishermen, Bekir told me another day, specialize in fishing octopus and squid in Izmir Bay and in lagoon fishing in Homa. Bekir's son

Figure 3. A pelican receiving a share of the fisherman's catch, 2015. Photo by the author.

and daughter-in-law had worked precarious seasonal jobs and had recently opened a fish takeaway, but they aspired for the next generation to achieve more secure employment working for the state. As fish are trapped in the mobile infrastructure of the lagoon seeking warmth and food, Bekir said while driving me to the suburban train station after a fishing outing, so too had he felt trapped in his factory job. Extending Bekir's comparison, both fish and fishermen were caught in yet another trap: the changing infrastructural bureaucracy of the lagoon.

Up to Code

Homa Lagoon is a lively delta infrastructure, made and remade through flows of water, labor, capital, and regulations, as well as through changing "environmental imaginaries,"[40] forming a layering of infrastructural visions on its changing landscapes. In the late nineteenth century, after city engineers redirected the Gediz River northward to prevent Izmir Bay from becoming too shallow, the Aegean Sea started eroding the lagoons.[41] The expansion of drainage and irrigation canals and the growth of urban districts in the twentieth century reduced freshwater flows. Lagoons became shallower, polluted, saltier, warmer, and eutrophic. One large Izmir lagoon was turned into a beach, then a trash landfill, and then a leafy park. In the 1990s, the municipality dredged the inner-bay lagoons.

The remaining lagoons were caught in contestations over access to infrastructure and ecological change. In the 1980s, as ornithologists began mobilizing against the expansion of the Çamaltı Saltworks, the university set out to acquire Homa with the aim of building a scientific fishery. The Department of Fisheries and Aquaculture won a long legal struggle against the fishing cooperative, and Homa became the only university-owned lagoon in Turkey.[42] Members of the cooperative could keep fishing, but the university would retain a large share of the catch. In addition, they would take over the infrastructural maintenance. Fishermen, however, told me that university staff would not do the seasonal work in the lagoon as they had done—a neglect that fishermen contended had accelerated the deterioration of the lagoon ecosystem.

In December 2009, a winter storm destroyed the access road. The university placed a locked gate at the lagoon entrance. Passes were distributed to "authorized fishermen" traveling to the docks beyond the gate, but the lagoon remained off-limits. Eating fried fish in his hole-in-the-wall restaurant, Bekir's son told me that he once tried to get in using his father's access pass and was subsequently banned from the lagoon. In our conversations in the university department in the winter of 2014, the lagoon managers (university aquaculture professors) explained that they closed access to renew the lagoon's infrastructure—the research building, road, barrier islands, and the fish traps—and to prevent the fisheries' depletion.

During a conversation we had in his office in spring 2014, the department head and lagoon manager characterized the fishermen then fighting to regain access to Homa Lagoon as *cahil*, "ignorant," and *güvenilmez*, "untrustworthy." His colleague declared, without irony, that the university had "brought the lagoon up to code." He pulled out a booklet on the legislation concerning a fishing lagoon's steel bar spacing and gave me another copy to keep. Seeking to construct stable infrastructures in unruly, unpredictable environments, experts reproduce their own notions of politics.[43] In this case, university lagoon managers made an argument about the positive ecological effects of their new and "up-to-code" infrastructure, invoking national legislation to justify restricted access and naturalize their classed distrust of the lagoon fishermen.

Private security guarded the lagoon. They spent their days watching TV in the damp and dilapidated building, making tea and cooking simple meals on a gas stove. At night, a guard told me, there had been intrusions by "illegal" users, sometimes armed. Unable to access the lagoon with the banned fishermen, I accompanied a university microbiologist on a dinghy as he fished for valuable organisms, filtering water with a makeshift tool of cheesecloth, water bottles, and a bullet cartridge. Back in the lab, the head of his lab showed me the phytoplankton through a microscope, describing its potentially lucrative industrial applications.[44]

In 2010, the municipality restored the lagoon sandbar with truckloads of rubble from demolished informal neighborhoods that had been constructed around the ancient Agora, including pieces of the Agora itself. The university, sued by the provincial government, had to pay a hefty fine, remove the rubble, and rebuild the sandbar.[45] The DSI installed a pump to allow seawater to enter the lagoon, as the inlets had been closed by the new road. In 2015, the university's vision for lagoon restoration entailed using mud extracted from the dredging of the bay to make new islands for nesting pelicans. It also involved redirecting water from wastewater plants into the lagoon through an underground canal. The project included a circulation canal dredged at the bottom of the lagoon—the first in the world, university scientists boasted.

When the university did not allow fishermen back into the lagoon, Bekir penned letters to state officials and institutions and held meetings and press conferences. His petitions were to no avail. In our conversations, he and other fishermen repeatedly described to me how the renovation project had radically changed the infrastructure of fishing and flow. The work that the fishers had performed seasonally to maintain water circulation and fish populations—opening and closing the inlets between lagoon and sea, made of reeds and rocks—was replaced with less mobile steel, cement, and gravel. The renovation works and a new management model aimed to create a new moral ecology as a model ecosystem:[46] a reengineered wetland laboratory and living space for university scientists, birds, and microorganisms.

This transformation was not just a word game: the wetland denomination also defined who (humans and nonhumans alike) could shape its material and symbolic futures. This argument was fought through envisioning new infra-structural transformation to produce different (and differently inhabitable) ecologies, all of which carried competing moral stakes for fishermen, scientists, residents, and environmental advocates. The following sections describe the enrollment of delta birds, particularly flamingos, in urban middle-class visions of moral ecologies of deltaic infrastructures against large-scale infra-structural transformations in and of the wetlands.

A Funeral for the Flamingos

Moral ecologies can be strategic for bringing together contradictory visions of infrastructural livelihoods. In September 2013, Deniz and I were drinking cups of filter coffee in the bookstore near my house, in the neighborhood of Alsancak in Izmir, and snacking on sesame cookies. Deniz told me about a new middle-class community uprising, right at the urban edge of the Gediz Delta wetlands. On a Friday in early October 2013, a small group of middle-class residents of the seafront neighborhood Mavişehir staged a "funeral for the flamingos." Carrying two flamingo-sized cardboard coffins onto which they had taped photographs of flamingos, the crowd walked from a residential complex, where most of them lived, to the pedestrian path alongside the Ramsar boundary of the wetland, which was marked with a signpost. The protesters had taped manifestos on the signpost: "Ecological destruction and environmental massacre at our doors: Are you aware?"

They walked in front of a large empty lot, overgrown with grasses. The lot, located within the Ramsar boundary, had been rezoned from "sea" to "land" in the latest revision of the zoning plan two years earlier. Residents of the high apartment towers nearby had learned about the zoning change and the subsequent plan for the construction of new high-rise complexes. They had filed several lawsuits to city authorities to halt the development, to no effect. Subsequently, they tried to attract media attention by staging a symbolic funeral for the flamingos. The flamingo photographs taped on the coffins were marked with the birds' call for help: "This is our house, not yours. Heeeelp!" Seven days after burying the flamingo coffins in the waters of the wetland, the residents distributed fried *lokma* sweets to passersby for the soul of the dead birds—following the local Sunni Muslim mortuary tradition.[47]

This multispecies burial drew on entangled symbols of state and religion that ultimately claim sovereignty over subject bodies, extending it to the wetland birds to recast them as political and moral subjects. The human participants constituted themselves as kin to the birds, in similar ways to how people performing the prescribed mortuary ritual practices of cleansing, prayer, burial, and mourning publicly reaffirm themselves as "family." Transgressive lives require transgressive mortuary practices, as anthropologist Aslı Zengin has theorized in her work with transgender sex workers in Istanbul, and may engender new forms of intimacy.[48] The flamingo coffins were to be buried in the wetland, thereby recasting the shallow waters as a place of death rather than one of thriving life and biodiversity.

In this case, however, the residents were not simply transgressing species boundaries in a shared performance of anticipatory grief for the lost wetlands. Rather, they were drawing on a rising Turkish political tradition of using symbolic burials as powerful performances of protest. For example, in 2014, Syrian refugee students symbolically buried Syrian war victims; in 2016, Aegean farmers performed burials of their grapes against decreasing market value; and indebted contractors during the 2018 economic crisis have buried "public contracting."[49] These public funerary performances all leveraged the symbolic social form of the burial ritual to call attention to a community's weakening and destruction through the ritual burial of their crops, their trade, their ethnicity. In contrast, the Izmir protesters' funeral was literally staged on behalf of the birds. It allied a moral ecology of bird livelihood in the wetland with state-sanctioned Muslim rituals of belonging and sovereignty, thereby constituting the birds as political and religious subjects.

I did not immediately connect this small mobilization to the Gezi Park protests in Istanbul. It had been just over four months since thousands of Istanbul residents took to the streets protesting authorities' attempt to transform a park in the center of Istanbul into a commercial complex. Met with violent police repression, the protesters kept pouring into the streets and squares of central Istanbul. The mobilization quickly spread to other cities, alongside countless grassroots forums on themes of democracy and ecology.[50] For my friends and interlocutors in Turkey, and for Turkish and foreign media commentators alike, this was also a time of renewed democratic hope and creativity, connecting ecological concerns with claims for livelihood, community sovereignty, and democracy. Participants in the protests and in the grassroots initiatives that sprouted in its aftermath often stated they had opened a new arena for such claim making, reaching beyond and across party and class boundaries. This moment of democratic hope and struggle endured despite the increasing political repression that followed the protests in the subsequent months and years, persecuting the movement leaders and its participants.[51] In the following years, political discourses predicated on idioms of "hope" continued to characterize the landscape of dissent in Turkey.[52]

Three weeks after the flamingo funeral, the residents invited Deniz for a meeting, which I also attended. At the meeting, they asked him to explain to the group the ecological functions of "wetlands" and the meaning of the different wetland conservation denominations in the country. Deniz and other Doğa Dergeği members, including me, organized a bird-watching event for the residents the following Sunday. Equipped with binoculars and birding books, they taught a group of about fifteen residents to recognize the wetland birds in the marshes just outside their apartment.

We posed for a photograph at the Ramsar wetland sign in the planned construction site (which in fact indicates the boundary of the buffer area), where protesters had taped leaflets three weeks before. The residents all donned clean gym shoes and leisurewear, the attire that these wealthy urbanites would wear on a short walk or jog, while Deniz, his collaborators, and I wore more "outdoorsy" hiking outfits and boots. A couple of residents had brought their cameras and took photographs of the flamingos feeding in the marshes on a background of high-rises and city haze.

Gökhan and his wife, Sema, two retired doctors living in a luxury high-rise, joined their neighbors in staging the legal struggle to stop urban expansion on the adjacent coastal wetlands. In the process, after consulting cadastral maps and historical zoning plats, the couple came to realize that their residential complex had itself been built on the Gediz Delta's wetland. Ironically, their apartment tower was called Flamingo Apartments. They recognized that the opposition to the new construction had initially stemmed from residents with sea-facing balconies concerned with maintaining their view. But what they learned during the months of nationwide mobilization had generated their awareness of the wetland ecosystem as holding livelihood rights—a moral ecology of urban life.

After the Gezi Park mobilizations, Gökhan, and Sema told me one day in the winter of 2014, there has been more public discourse on issues of livelihood and environmental protection.

"I grew up and lived in a village," Gökhan explained, while Sema went to the kitchen to pour us cups of tea. "I took for granted the environment in which we lived: the trees, meadows, fields. As we started to build everywhere, we realized that no environment would be left behind. We destroyed so much of the environment that we realized its value."

"What does the word 'environment' mean to you?" I asked.

"Environment," he said, pausing for a few seconds, "is the place I inhabit. It is also the place birds inhabit. The place trees inhabit." Talking about inhabiting, in the sense of living and making a home, the ability to thrive in one's livelihood as rooted in space and community, Oğuz articulated a moral ecology that conjoined the built environment and ecological relations. Here Oğuz emphasized the made quality of the wetland, and that, while some infrastructures are welcoming to specific forms of multispecies life, others lead to death.

The organizing efforts, however, proved fruitless. Three years later, two towers almost thirty stories high were built at the edge of the wetland. One, the Port Marin complex, featured luxury three- and four-bedroom apartments, advertised by the construction company in digital images of spacious living rooms featuring tropical oceanic vistas—not the shallow wetlands of Izmir Bay. Critical commentators have dismissed the ecological motivation of the protesters: this was merely a middle-class defense of the expensive seaview balconies. Yet, while these more prosaic concerns of real estate value and aesthetics certainly had also contributed to the uprising, at least for the owners of sea-facing condos, many residents, such as Gökhan and Sema, learned to make legal arguments leveraging Ramsar and national conservation statuses against the expansion of urban development on the wetland. In the process, they appropriated categories of the wetland and of wetland services as their own and, inspired by the Gezi mobilization, expanded their own concern of livelihoods to the livelihood rights of birds, plants, fish, and microbial life.

"Don't Touch My Flamingo"

Well after the hope arising in the aftermath of the Gezi mobilization had given rise to fear of antidemocratic repression, rising authoritarianism, and persecution, particularly after the coup attempt of July 2016, moral ecologies of the wetland continued to be expressed through civil-society organizing and lawsuits against government bodies. On May 4, 2017, three civil-society organizations and eighty-five citizens filed a lawsuit in Izmir's district court against the Ministry of Environment and Urbanization. The 2014 revision of Turkey's wetland legislation had eased new infrastructure development in protected areas.[53] This coalition was now contesting the ensuing approval of the Environmental Impacts Assessment for a highway bridge over Izmir Bay. In February 2017, my friend Zeynep told me, the National Wetland Commission had changed the Gediz Delta's conservation boundary near the area scheduled for bridge construction from a "strict preservation zone" to an "area of controlled use."[54] This had simplified the approval process for infrastructural development; in August, the NGOs filed another lawsuit against the new conservation zoning.

The contested project consisted of an eleven-kilometer-long highway, including a bridge, a tunnel, and an artificial island in the shape of a moon and star. These were the emblems found on Turkey's flag, but friends told me that the island would be made in the shape of a light bulb, the symbol of Turkey's ruling party. The rumor about the "real" shape of the island suggested that the project had come to stand for the ruling party both as its material incarnation in the environment and as an authoritarian mechanism for the exercise of centralized power. The bridge would cross from the southern bay to its northern shores in the Gediz Delta and then connect to the new Izmir-Istanbul highway. The project had been at the forefront of AKP politician Binali Yıldırım's losing municipal election campaigns in 2011 and 2014.[55] Its proponents referred to the project as a "necklace" for the city, and the bridge was named as one of the AKP's goals for the republic's centenary in 2023.[56]

Zeynep, a civil-society organizer and environmental social scientist, emailed me the lawsuit document. I read it as a moral ecology couched in the language of law to protect human and nonhuman livelihoods, and Zeynep agreed with my analysis. Infrastructure mediates state and corporate power, collective actions, and subjectivities—often becoming visible when things do not work as planned. For instance, failing or malfunctioning urban water infrastructure reveals political systems that leave their citizens to patch up broken systems by themselves,[57] and it can give rise to new gendered practices of care to maintain family and neighborhood ties while dealing with overflowing sewage, ill-fitting connections, and unpredictable water bills.[58] While access to water infrastructure may be central to communities' claims of belonging,[59] engineers who design new infrastructures for water and electricity provisions also reenvision users as (im)moral subjects.[60] In this case, and in the example of the residential complex, Izmir residents formed new alliances mediated by a shared concern with an unwanted infrastructure and expressed this concern through idioms of ecological connections and bird livelihoods.

The bridge, the plaintiffs declared, citing Izmir's latest urban plan, was incongruous with the city's transportation needs. Yet most of the opposition to the bridge was not couched in economic or planning arguments but in ones of ecology and livelihood. The bridge project, poised to increase water pollution in Izmir Bay, would result in the destruction of the ecological life and biodiversity of the Gediz Delta's wetlands. In contesting the bridge, plaintiffs leveraged the international Ramsar status of the delta's wetlands alongside other national nature conservation statutes.

The lawsuit emphasized the role of the delta as a feeding and nesting ground for hundreds of bird species, including one-third of the Mediterranean flamingo population.[61] In the wake of the June 2017 court decision to reassess the Environmental Impact Assessment (EIA), Zeynep and other friends staged a social media campaign centered on two connected slogans: *flamingoma dokuma*, "Don't touch my flamingo," and *Gediz Deltası hepimizindir*, "The Gediz delta belongs to all of us." In a press conference I attended in June 2017, the environmental NGO's spokesperson declared that the project would "cause one of the biggest destructions ever seen in world history."[62]

The flamingo, a mascot of wetland conservation in Izmir—and tangible proof of its putative failures and successes—had become a charismatic symbol of the varied nonhuman livelihoods potentially destroyed by the planned project. Yet it was not simply a charismatic animal leveraged instrumentally.[63] At stake in the stated concerns for ecological life were also deep-seated, emotional, cultural, and ethical concerns with (human) power and with the effects of the authoritarian planning of infrastructure. These were questions of infrastructural moral ecology. In July and August 2018, a court-appointed panel of experts stopped the project for its lack of compliance with environmental protection regulation in the wetlands.[64]

In October, the court held that the project's EIA was invalid, stating its negative impacts "on a very important nature conservation area, protected by international environmental conservation agreements as well as by other protection statuses" and "on water temperature and circulation in the bay, which would result in the disappearance of *Artemia* [brine shrimp], which are fundamental to the food chain of flamingos, and that would also cause damage to fragile ecosystem equilibria."[65] The general elections of June 24, 2018, which resulted in the victory of the People's Alliance, an electoral alliance formed between the AKP and the nationalist MHP (Milliyetçi Hareket Partisi; Nationalist Movement Party), ushered in sweeping constitutional change and an authoritarian presidency—soon followed by the rapid fall of the Turkish lira. In its aftermath, less media attention was devoted to the bridge. But for Zeynep and others, this was a glimmer of hope and resilience.

Leveraging Ramsar

In the wake of the 2017 anti-bridge campaign, I emailed Thobias Salathé, the senior adviser for Europe at the Ramsar office. I wanted to know, from his perspective, whether the Ramsar office could provide more than symbolic political leverage for the anti-bridge activists. I had first met Salathé at Turkey's National Wetland Conference held at the University in Samsun in October 2013, where he had given the conference keynote address, "Wetlands Take Care of Water." In 2014, I interviewed him during my visit to the Ramsar offices in the Swiss town of Gland. Salathé had been at the Ramsar office for more than two decades, after working at the ICBP and at the Tour du Valat's wetland research center.

I asked Salathé how the Ramsar office had responded when the Doğa Derneği director notified them of the lawsuit against the bridge. Salathé's answer exemplified how Ramsar office staff positioned themselves as technical consultants, deferring to national bureaucracies. When a wetland is threatened by planned development, Salathé summarized for me, the Ramsar Convention (Article 3.2) redirects to a ministry appointed as the Ramsar Administrative Authority (AA). At the time of our email exchange, the AA in Turkey was the Ministry of Water and Forestry. If the AA agrees that the planned development is "likely to influence the ecological character of the wetland," the Ramsar office might propose an "advisory mission" on-site to produce a document with recommendation. Alternatively, the AA can ask the Ramsar site to be put on the "Montreux Record," which is a list of wetlands threatened by "adverse change in ecological character," a process that involves more thorough documentation.[66] However, Turkey has never filed a Montreux request, and the AA did not summon the Ramsar advisory mission to investigate the bridge development in the Gediz Wetlands.

Ramsar office staff rarely traveled to the Gediz Delta or other Ramsar wetland sites, so they could not intervene directly in the debate on the construction of the highway bridge over the protected wetlands. Rather, Ramsar sought to maintain a global outlook, for instance, maintaining a digital worldwide database of Ramsar wetlands; hosting the triennial conferences for

hev how
the
reader
sees the
last
material

the participating countries; and publishing educational material, technical reports, and "best practices" manuals. Even though the Ramsar status held no actual legislative enforcement in Turkey, the denomination of the Gediz Delta as a wetland of "international importance" became a toolkit that activists and citizens used strategically in their moral claims about infrastructural transformations. They leveraged the Ramsar status to bolster their argument about the moral ecologies of the delta—human and nonhuman livelihoods that would be disrupted by the bridge construction.

The international status of the Gediz Delta wetland, then, acquired a political significance at the regional and national levels because it could be used to present arguments that, discursively, were beyond the divisions of party politics and perhaps more sheltered from the risk of political repression. Yet the actual unfolding of wetland contestations remained enmeshed in local political processes, as this chapter demonstrates. The Ramsar denomination, having produced the wetland as an object of environmental governance, care, and struggle, remained less effective as a political tool than locality-based contestations motivated by specific moral ecologies.

Conclusion

Izmir residents, fishermen, NGO workers, scientists, and bureaucrats understand the Gediz Delta as a livable place by invoking infrastructural remakings of specific relations. Infrastructures are ecological, inhabited and produced through work, power, and capital. Infrastructures, I contend, exist in varied relations to environments and ecological relations and in ways that are not merely metaphorical. Given its broad analytical purchase, it is unsurprising that anthropologists see infrastructure as at once "generative and degenerative; constructive and destructive; future oriented but ultimately fleeting."[67] In the Gediz Delta, these apparent contradictions (inflexible and mobile, for instance, or futuristic and traditional) arise as scientists, state officials, and residents construe varied moral notions of ecological relations and livelihoods. Perhaps the Gediz Delta is an "enchanted infrastructure," creating a sense of shared social good and holding together competing hopes,[68] and this enchantment involves ecological relations alongside social or political ones.

This chapter foregrounds wetlands *as* ecological infrastructures, which exist within and, in turn, shape lived environments of human and more-than-human ecologies. These infrastructural wetland environments are suffused by contrasting moral understandings of normative relations and justice. Ecologies come to matter as moral landscapes always already embedded in infrastructure. Disentangling infrastructure and ecology as separate categories can be accomplished only by erasing a variety of perspectives. For example, the fishermen's cooperative argued that their infrastructural labor created lagoon ecologies that were viable for sustainable fishing. In contrast, university management contended that only their infrastructural renovations, "up to code," would create a valuable lagoon ecology for scientific knowledge production. Similarly, some university professors and state officials envisioned the role of irrigation infrastructure as sustaining bird ecologies and posited the delta's eucalyptus trees as invasive organisms to be removed. The outcome of the co-constitution of infrastructure and ecology is a moral value, in E. P. Thompson's sense of a shared sentiment.[69]

A moral ecology is both an impulse for action, Emre's mapmaking, for instance, and the terms in which claims are made, such as his vision for ideal relations between birds, fish, flows, and infrastructure. Moral ecologies are also couched in epistemological claims: Zeynep, Sıkı, Emre, and the Flamingo Towers residents all claimed a scientific understanding of delta ecologies, appealing to different, and often contrasting, scientific arguments tempered with moral commitments to "good" landscapes. And Bekir and Ahmet leveraged their experiential command of deltaic flows, one gained through decades of lagoon making, fishing, and seafaring labor. All of them also practice moral visions of ecology of infrastructure as they create and sustain social relations with the delta's pelicans, flamingos, *Salicornia*, reeds, fish, and countless other nonhuman beings while working, foraging, conducting research, spending time with family, and cultivating friendships and collaborations.

4 Caring for the Delta

Learn to Feel the Wetland

One warm September afternoon in 2014, Mustafa, a professor of agricultural engineering at a university in Samsun, convenes a group of college students in the Kızılırmak Delta wetland conservation area on Turkey's Black Sea coast. Climbing on the squeaky wooden steps of the bird-watching tower at the visitors' center, one can gaze across the delta's expanse of wet meadows and shallow lakes and lagoons and see the reed beds, swamp forests, and sand dunes stretching out to the Black Sea coast. Herds of water buffaloes, sheep, and horses graze in the common pastures, guarded by lone shepherds riding their horses or motorcycles on the meadows. Beyond the boundaries of the conservation area, marked only by official signs positioned on the one main road, fields of rice and corn are almost ripe for harvesting.[1]

Behind Cernek Lake, the largest wetland in the delta, hills and mountains are obscured by light gray clouds. Closer to the wetland area's management center, canals and a water pump station suggest the presence of the extensive network of irrigation that wets the rice fields, and the drainage canals that, in the second half of the twentieth century, turned the coastal delta into productive agricultural land. Not visible from this tower, though central agents in the making and remaking of the delta's form and ecology, are the large dams upstream on the Kızılırmak River and, downstream, fifteen kilometers from us, the bustling commercial town of Bafra.

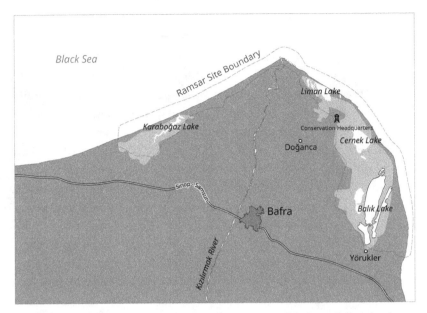

Map 3. Kızılırmak Delta, showing the main sites mentioned in the text. Drawing by Benjamin Siegel.

"Take your shoes off," Mustafa commands, and gestures toward the shallow water ahead. Thirty young men and women and I start off barefoot toward the soft shores of Cernek Lake. Thorny plants prick our feet until the slippery mud grows deeper. One student falls into the mud but trudges forward gamely, her jeans now caked in gray. Water buffaloes turn apprehensively toward us: some lumber away from the pools of mud, water trickling off their furry bodies, while others seem blithely indifferent to our presence. As we circle back toward the fenced lawn and paved ground of the visitors' center, the students relax, chatting animatedly on firmer soil. "Now," Mustafa says with a smile, "you have all learned to be water buffaloes—*to feel the wetland.*"

This chapter analyzes the transformation of the Kızılırmak Delta's fields, marshes, swamps, meadows, canals, lakes, and reed beds into a conservation wetland. This production *materialized* international categories of wetlands and national imaginaries of ecological value and care into land, water, plants, and animals. Over the course of the twentieth century, population and agro-economic shifts concurrent with the formation of the Turkish nation-state transformed the Kızılırmak Delta into a thickly inhabited and productive agricultural landscape. Subsequently, starting in the 1980s, the coastal regions of this agrarian delta were recast as a conservation area, informed by changing Turkish and international notions of wetlands. Through local processes and everyday practices, such as Mustafa's exercise, the Kızılırmak Delta wetlands become relevant to different social groups as subjects of scientific knowledge, ecological care, and environmental imaginations. These practices have rendered the wetland an open-air laboratory and an object of care for environmental advocates, scientists, and residents. The delta's wetlands have become grounds for contestation between different social groups, as well as for an unexpected and contingent flourishing of friendships and alliances. Knowledge, care, and imagination are all entangled in manifestations of moral ecology and its contested politics.

By the time Mustafa's students stepped into the wetland lake, I had been living in a rural village in the Kızılırmak Delta for four months in a farmhouse three kilometers from the wetland conservation areas' management headquarters. I wanted to understand how wetland conservation had become entangled in farmers' everyday moral ecologies, work practices, and contestation over access to land and water. In another delta conservation area, the Gediz Delta, on Turkey's Aegean coast, I had studied the everyday work of an environmental NGO staff, city bureaucrats, university scientists, and fishermen. The Gediz and Kızılırmak Deltas had become a protected Turkish wetland beginning in the late 1980s thanks to the advocacy work of local environmental scientists and organizations. Environmentalists, scientists, and bureaucrats looked at the two deltas to compare and assess models of conservation management, civil-society participation, and anthropogenic threats to wetland biodiversity. Yet each delta was characterized by unique dynamics and processes, inflecting local meanings of the wetland denomination and generating varied contestations around ecological livelihood.

In October 2013, I attended Turkey's National Wetland Conference, hosted at the university in Samsun. After the conference, Mustafa—one of the conference organizers—introduced me to farmers in the village of Doğanca, one of the largest rural communities in the delta that borders the conservation area. One family of smallholders offered to host me on the upper floor of their house. With them, and with the other families in Doğanca and other delta towns who welcomed and hosted me, I participated in the everyday life and work of this agrarian community. On a typical day, I woke up at five in the morning and helped my hosts tend to their water buffaloes or vegetable gardens. I fed the chickens, gathered eggs, and set up the breakfast table. Often the women cooking breakfast would task me with calling family members and hired workers to the kitchen. After a breakfast of copious tea, buffalo cream, eggs, jams, cheese, fried vegetables, and white bread from the bakery in town, I continued the farm work for a few more hours before returning home to help cook lunch.

One week, one of my hosts tried to teach me to drive his tractor, an effort that ended when I almost drove both of us into a canal. My work on the farm was the kind of labor usually performed by women, and I did not work in the large cash-crop fields, which would require a familiarity with tractors and with handling heavy farm machinery. The days I spent on small dinghies with local fishermen were an exception to the gendered spatiality of the wetland, where men control hunting and fishing activities, although poorer women participated in harvesting and cutting reeds in the lake alongside their male relatives. In the afternoons, I visited other families in the village, followed my hosts to run errands in town, or joined wetland scientists in various meetings, fieldwork activities, or education trips.

The Kızılırmak Delta farmers have themselves become objects of ecological care, as environmental conservationists have enrolled, romanticized, and excluded agrarian livelihoods in their visions of the wetland. In the two years I spent in the delta,[2] I found that as local scientists and advocates had transformed the agrarian delta into a wetland, the new denomination encompassed the productive activities of delta farmers. Wetland conservationists sought to involve the farmers as drivers of biodiversity and repositories of traditional knowledge, writing out the history of settlement in the delta as well as contemporary farmers' perspectives and their aspirations for work, class mobility, leisure, and sociality.

Crafting Relationships of Care

Later in the evening, sipping glasses of tea in Bafra, I speak with Mustafa about his exercise. A well-liked local specialist in organic agriculture and wetland management, Mustafa has been leading a wetland education camp for university students. This undertaking, funded by Turkey's Scientific and Technological Research Council, is coordinated by local university professors, and I have joined the wetland school as a volunteer assistant. Mustafa is working to instill in his students similar questions to the one that brought me to the Kızılırmak Delta two years earlier: How are wetlands produced, discursively and materially, as wetlands?[3]

Over the course of the twentieth century, the swamps and marshes of the lower Kızılırmak Delta have undergone two major transformations. First, they became a productive agricultural region enmeshed in national and international markets—a transformation powered by the labor of resettled populations from former regions of the Ottoman Empire and by rural migrants from poorer Turkish locales. Then they came to be seen as a wetland ecology to be protected from further agricultural expansion and development. Like Mustafa, I understand these two transformations as entangled. Unlike me, Mustafa has also tasked himself with producing the Kızılırmak Delta as a valuable wetland through the exercise of scientific knowledge making, a process that takes place in scientific papers, in political advocacy, or in asking students to step into a lake.

Leaning back, Mustafa reflects on the importance of "feeling" the wetland landscape—of identifying sensorially with mud, buffaloes, water, plants, and thorns. Yet sensing, or feeling (hissetmek), is not enough. After students come to feel part of the delta, he explains, they begin to pay attention to the wider processes impacting it: from conservation policy, to pollution, climate, infrastructure, cultural practices, and farming practices. He describes this work as fark ettirmek, meaning "to raise awareness" or "to make someone notice," which he connects explicitly to physically being in place and paying attention. Mustafa tells me that he understands the Kızılırmak Delta as a valuable yet threatened ecology. He explains to me that recent changes in infrastructure, water flows, pollution levels, and agricultural practices are leading to both the loss of wetland habitats and possibilities for ecological life.

Mustafa, his colleagues, and his students are crafting relationships of care—political, scientific, cultural, material—that create and sustain ecologies for humans and nonhumans to inhabit. These care relations are driven by emotional and ethical commitments, overriding utilitarian concerns, and do not preexist these social, bodily, and sensorial knowledge practices.[4] Scholars of Turkey have recently highlighted the importance of attending to the changing politics of expertise in biomedicine, in the construction of large infrastructural projects, and in seed banks and agro-businesses.[5] This work foregrounds the complex social and political lives of engineers, scientists, and bureaucrats and the affective and cultural motivations underpinning techno-scientific work. For the scientists and environmental advocates I worked with, wetland transformations raise a set of vital and moral questions. What kind of ecology should the delta-as-wetland be, and for whom? Who is excluded in such a configuration? Wetland advocates' own efforts to answer these questions evinced the resonance of wetland conservation, research, and education with agrarian moral ecologies. At a moment when Turkey has become known for widespread social turmoil, political repression, and environmental degradation, the work of wetland advocates demonstrates the ongoing and widespread salience of ecology as generative of social worlds, cultural meaning, and democratic politics. At the same time, farmers, and the environmental and economic precarity of agrarian work, continue to be romanticized and marginalized in emerging visions of ecological livelihood.

I propose that we attend closely to the care practices of wetland advocates. Anthropologists of Turkey have long situated practices of care at the heart of notions of morality, kinship, belief, and modernity.[6] More broadly, the analytic of care in anthropology has been widely employed to theorize kinship relations, notions of morality, and power asymmetries—and even to citizen's everyday access to subsidized staple foods.[7] Writing on farmers' ethical engagement with fields, glaciers, and domestic animals in the mountains of Ladakh, anthropologist Karine Gagné has recently argued for a notion of care as intimate, embodied knowledge and "practical engagement informed by a sense of obligation and responsibility towards the non-human."[8] Mustafa and other environmental conservationists in the wetland would agree with political scientist Joan Tronto's expansive understanding of care as "everything that we do to maintain, continue and repair 'our world' so that we can live in it as well as possible."[9] Imagination of livelihoods and speculation about possible futures (and value) are central to environmental experts, like Mustafa, who have come to care for ecologies in moral ways.[10]

Focusing on the care of the wetland to emphasize the daily practices, ethics, and relationships that inform the work of scientists like Mustafa, I foreground the affective commitment that undergirds an "expert" concern for degrading environments.[11] However, care is both productive and destructive: while care creates and sustains some relationships, it simultaneously damages others.[12] This is no less axiomatic in Turkey. Moral ecologies can be oppressive and destructive and can silence others. I propose that they must be understood in the context of historic and current remakings of environment.

Environmental conservation projects are entangled in the violence of colonial histories, and regional specificity matters. Nature conservation areas were established by colonial governments and postcolonial national elites, displacing subsistence populations and indigenous communities.[13] Conservation projects materialize nationalist and colonial historical narratives, whether by reconstructing ancient wildlife species or unearthing archeological remains.[14] In contemporary Turkey, concerns with environmental destruction and ecological loss are now closely tied to broader contestations about protecting rights to livelihood and democracy.[15] Stewarding a renewed ethnographic interest in the contested ecologies of Turkey, scholars have focused on cultural narratives of environmental degradation,[16] the micropolitics of irrigation infrastructure,[17] resistance to neoliberal valuations of water and energy,[18] the temporality of large-scale dam projects,[19] activist mobilizations against nuclear energy, and responses to neoliberal development paradigms and authoritarian governance.[20] In the Kızılırmak Delta, communities of environmental scientists and managers practice their moral commitment to care for ecologies, as they confront environmental and agrarian change and wrestle with the changing livelihood practices and aspirations of delta farmers.

Making Turkish Ecology in the Kızılırmak Delta

Pontic geographer Strabo (63 BCE–23 CE) described Gazelonitis, the delta of the Halys River (today's Kızılırmak) near his natal city of present-day Amasya, as a "fertile country, wholly consisting of plains, that produces every kind of fruit. It also affords pasture for flocks of sheep, which . . . produce a soft wool."[21] Yet littoral transformations have meant that this coastal region is no longer the land described in Strabo's *Geography*: contemporary geographers estimate that since the turn of the Common Era, the delta has expanded toward the sea at a rate of ten meters each year.[22]

At the turn of the nineteenth century, the delta was a maze of swamps, wet meadows, and large shallow lakes. Shepherds grazed sheep and water buffaloes, hunted, fished, and gathered reeds for thatching and weaving. On the higher land surrounding the lower delta, people grew hazelnuts and a New World crop: tobacco. First encountering the crop in the sixteenth century, the Ottoman Empire became the foremost exporter of tobacco to Europe by the eighteenth century. The growth of tobacco transformed markets and populations: after the formation of the Public Debt Administration in 1881, tobacco production in the Kızılırmak Delta came under the purview of a French company, the Régie, backed by European banks to alleviate Ottoman debt. Greeks and Armenians controlled tobacco production and trade.[23] After 1923, tobacco became a Turkish state monopoly. Newly resettled Muslim Greeks in the delta took over the work of tobacco cultivation, together with Roma communities. The delta grew more closely embedded in global cash-crop markets.[24]

Today, Turkish and international environmental advocates often talk about the Kızılırmak Delta's farmers as traditionally connected to the land. However, the delta was thoroughly resettled in the twentieth century. Circassians escaping from persecution in the Russian Empire in 1864 were offered the marshes of the Kızılırmak Delta instead of the higher mountainous lands they had desired; almost all perished from malaria during the first summer. Muslim populations migrated from the Balkans during the war of 1877–1878. Hundreds of people from Kosovo were resettled among delta villages during the Balkan Wars (1912–1913).[25] During the Committee of Union and Progress government (1913–1919), hundreds of thousands of Armenian and Greek men and women were killed and deported. Beginning in 1915, Ottoman authorities conscripted Armenian men in labor battalions, subsequently deporting Armenian populations from Anatolia to concentration camps in the Syrian desert via harrowing death marches.[26] Similarly, starting in 1914 and culminating in 1919–1921, Greek Orthodox communities were forcibly conscripted, deported, and killed.

In Bafra, Sinop, Samsun, and surrounding areas, a series of armed operations in the winter of 1916 and an intensification of executions and deportations in 1919–1921 resulted in the violent suppression of Greek and Armenian civilians.[27] Local memories of the Greek and Armenian communities and their material traces in the land—houses, fields, churches, schools, fountains, cemeteries—have subsequently been elided.[28] In the wake of the Lausanne Treaty of 1923, which defined the boundaries of the Turkish republic, the remaining Greek Orthodox residents of Anatolia were forced to leave. Muslims from Greece were resettled in Turkey, often in the same villages and houses that had been expropriated from Greeks and Armenians. The treaty resulted in about two million people resettled on both sides.[29]

By 1926, the Kızılırmak Delta had fifty-six thousand residents, six thousand of them Muslims resettled from Greece after 1923.[30] From the 1940s onward, more people migrated or were formally resettled from Albania and Bulgaria. The resettlement of Muslim populations from former Ottoman provinces was concurrent with increased drainage of wetland to create new agricultural land and settlements. Imperatives of wetland drainage and

reclamation thus grew closely sutured to projects of Turkish nation formation. Wetland drainage intertwined visions of national development with
concerns over public health and political stability. In 1926, the Turkish government passed legislation to combat malaria; its approach for the eradication
of malaria-carrying mosquitoes involved large-scale drainage of uncultivated
marshes and swamps.[31] Like their European and North American counter-
parts, Turkish officials viewed marshes and swamps (*bataklık*) as unproductive and unhealthy areas to transform into agricultural land. This sentiment
was amplified by a new strain of malaria entrenched throughout Turkey
through the expansion of rice cultivation.[32] Prevailing questions of livability
and civilization deepened conceptions of marshes as unhealthy places.[33]

Wetlands were drained not only to combat malaria or expand agricultural
production: the creation of new settlements coincided with the remaking
of ethnonational Turkish communities. Newly reclaimed coastal deltas and
inland marshes, transformed into areas of agricultural production, attracted
poor and landless peasants from other regions. Beginning in the 1950s, hundreds of peasants migrated to the Kızılırmak Delta, joining communities
that had been resettled from the Balkans. These migrants worked as herders,
workers, and sharecroppers for large-scale landowners. The lower delta population grew from 8,500 in 1930 to 43,500 in the 1990s.[34]

The deltaic environments of seasonal grazing, hunting, and fishing in
the marshes and swamps had changed rapidly during the first decades of the
Turkish republic. The passage of the 1950 Law of Drained Wetlands and Reclamation boosted state-led projects of drainage, delineating how newly drained
land should be redistributed to farmers.[35] After the 1953 founding of the DSI,
which was modeled on the US Army Corps of Engineers and its Reclamation
Authority, this new agency became responsible for all water management and
infrastructure projects, including wetlands.

Canals built under the aegis of the DSI continued to drain and irrigate
the Kızılırmak Delta wetlands, allowing for the expansion of cash-crop agriculture and irrigation for year-round cultivation. These canals transformed
ecologies and economies. Agricultural and urban wastewater was redirected
to the shallow wetland lakes near the coasts. By 2006, wet rice agriculture,
introduced in the 1980s, accounted for half of the delta's irrigated agricultural
land.[36] Rice demands larger intakes of water. Its runoff, rich with chemical
pesticides, herbicides, and fertilizers, has affected wetland plants and animals
in dramatic ways.

An Open-Air Laboratory

In the mid-twentieth century, the scientific category "wetlands" was established to describe the ecological value of countless marshes, bogs, swamps, and other places saturated in water that were being drained to become agricultural urban or industrial land. In the 1960s, Turkish scientists, bureaucrats, and environmentalists put this category to work for shallow watery ecologies, from lakes to lagoons, deltas, and wet meadows, to be salvaged as conservation zones. The Kızılırmak Delta was soon caught between imperatives of agricultural development and those of wetland conservation. In the late 1970s, Turkish state planning institutions and local environmental organizations began to plan for the creation of a nature park in the delta—resulting in comprehensive plans for expanding drainage and irrigation infrastructure and construction of several large dams on the Kızılırmak River.[37] Yet, responding to pressure from rural communities who had heard of the restrictive measures implemented in the other newly established conservation areas, authorities decreased the geographical extent of the newly established nature reserve.[38]

Beginning in the late 1980s, Turkish and European scientists, NGO workers, and bureaucrats in Samsun, a provincial capital of half a million residents, and in Bafra, a bustling rural municipality, came to reimagine the Kızılırmak Delta as a natural ecology. The delta was then a thickly inhabited and fertile alluvial plain, home to tobacco, melon, pepper, and cabbage fields; rice paddies; and herds of livestock grazing in the lower plains. The coastal marshes of the delta, long seen as unproductive, treacherous, and unhealthy, were newly cast as a font of biological and cultural value, at risk of disappearing in the face of urban, industrial, and agricultural pressures.[39] In 1994, as Turkey joined the Ramsar Convention, the Kızılırmak Delta became a SIT Alanı—a status that protects natural or cultural areas from construction. Two years later, the Ministry of Agriculture and Settlement and the Ministry of Environment jointly developed a comprehensive plan to define conservation areas and regulate land use. On paper, conservationists were following a classic model of "participatory conservation." However, rather than allow farmers to participate in environmental decision-making and set the conservation agenda, academics and experts would *teach* rural residents about sustainable practices. Rural residents continued to resist the restrictions on their livelihood activities implemented in subsequent conservation regimes.

In 1998, the coastal areas of the Kızılırmak Delta became a Turkish Ramsar site. According to a booklet published by the Ministry of Environment, the delta was an ideal "open laboratory for scientific studies." The author also highlighted the contribution of the delta's fisheries, reed cutting, livestock grazing, and recreational activities (hunting and bird-watching) to the regional economy.[40] This economic characterization of the wetland drew from the new concept of ecosystem services—the benefits, in dollar terms, wrought of natural ecosystems.[41]

As scientists, bureaucrats, and city dwellers have turned their gaze to the wetland, the twenty-five thousand rural farmers living in the delta have simultaneously grappled with these new denominations of value. While they have largely ignored the formal denomination "wetland," which has never replaced local names for coastal ecologies and lived places, farmers have occasionally appropriated the wetland category to their advantage, as later sections demonstrate. However, these remained marginal to scientists' and environmentalists' moral notions of ecology and care, even as they frame the moral value of rural livelihoods and traditional wisdom.

Contested Care

On an August afternoon in 2012, Yağmur and I are driving from her laboratory at the University of Samsun to the university's ornithological research station in the Kızılırmak Delta. Yağmur has been conducting research in the Kızılırmak Delta for many years. She participated in writing the Turkish government's management plan for the delta. We follow the Black Sea coastline. Farmhouses are perched on the lush green hills. Tobacco fields yield to rice, peppers, tomatoes, leeks, sugar beets, and corn. We reach Engiz, a dense settlement of tall apartment buildings punctuated by hardware and supply stores, formally renamed Ondokuz Mayıs in 1961. The highway continues toward Bafra, but we take a right on an unpaved road toward the conservation area, passing through swamp forest surrounding the village of Yörükler.

Yörükler is a general term for nomadic people,[42] and today's residents claim their origins in pastoral communities from the uplands who settled in the coastal plains in the first half of the twentieth century. Over the last decade, the swamp—in Turkish, *subasar* or *longoz ormanı*—has become a favorite destination for urban nature lovers and natural scientists.[43] In the winter, trees are submerged in water, and in the spring, nature photographers armed with telephoto lenses flock to the forest to snap the blossoming water lilies and ranunculus (*düğün çiçeği*). Beyond its draw for photographers, the forest is a popular stop for nature education field trips.

Scientists, environmental advocates, and the National Parks officials with whom I visited the forest in subsequent years would invariably point out small forest clearings planted with melons. Environmentalists blamed this patchy deforestation, symbolized by the offending melons, on villagers and on the failure of local authorities to enforce conservation restrictions. When I interviewed them in August 2012, three families of farmers in Yörükler spoke of their clandestine planting as deriving variously from economic need, increasing land scarcity, complicated accounts of heredity, land tenure, and aspirations of social mobility.

Yet, on a broader level, farmers are also continuing a process begun in the twentieth century: the transformation of coastal wetland and swamps from collective hunting and fishing grounds and grazing pastures into private agricultural lands. This continuity is also evidenced in rural delta residents' gathering of wild species—reeds used for roof thatching, sharp-pointed rush (*goga*) for weaving and ornaments,[44] and leeches for medical use—sold to local middlemen and distributed to national and international markets.

Yağmur and I pass by a sandy stretch of coast, and she points to a neighborhood of beach houses built as vacation residences. These houses, she explains, were built on land sold by villagers holding customary rights to urban middle-class vacationers—a transaction without legal standing. Still, the new owners had successfully mobilized against a court order to tear the houses down. A few houses hung Turkish flags in their windows as a patriotic claim to legitimacy. State officials, farmers, scientists, and environmentalists repeatedly described to me the cluster of beach houses as an eyesore, a material manifestation of corruption, and a cause of ecological degradation.

Yörükler residents articulated different grievances. A retired migrant returnee from Germany told me in 2012 that authorities had prevented residents from building wooden infrastructures for ecotourism, the result of a former mayor's efforts to develop the coast as a commercial beach. Other Yörükler farmers complained about the lack of action against the illegal beach houses and about the construction of the park's visitors' center and other administrative buildings in stone and cement. Why, they asked, were they being prevented from constructing in more sustainable materials and benefiting from the influx of new visitors? In 2015, the beach houses would eventually be demolished, leaving behind small mounds of rubble that were soon overgrown by the local vegetation. Yet rural residents remain caught between competing national imperatives of agricultural expansion, tourist development, and environmental conservation.

As we continue to drive toward the ornithology research center, I expect Yağmur to recount adventurous stories of rare-bird sightings or field encounters with hunters and wildlife. Instead, she begins to talk about water and infrastructure. She describes the delta as a complex landscape in movement, characterized by different kinds of water—open water, freshwater, and semi-saline lakes, marsh vegetation, sand dunes, woodland, and irrigation. The construction of water dams upstream on the Kızılırmak River in the last decades, she explains, has stopped the flow of sediment in the river. As a consequence, the Black Sea has been eroding the deforested coastal strip; eventually, she says, the last remaining lakes will join with the sea.

Gesturing toward the lake on our left and at the drainage canal on our right, Yağmur speaks worriedly about agricultural runoff that had previously been draining into one of the wetland lakes now being redirected into the Black Sea itself. The adverse ecological effects of the drainage canal in question, posed against its agricultural advantages, would remain a heated subject of disagreement among residents, conservationists, scientists, and state officials. It is a complicated matter, with supporters and detractors within each group.

We park in front of the ornithological research station, a wood and concrete building at the edge of the lake, unpacking groceries and field supplies. A small group of young men and women, clad in colorful cotton T-shirts, jeans, and green rubber boots, greet us at the door. Twice a year, Yağmur and two colleagues move to the research station for nine weeks, bringing along student volunteers from all over Turkey. I joined the camp twice, in 2012 and in 2013, and visited subsequently while living with farmers in Doğanca. Some students are already amateur birders, and many joined simply out of a desire to experience the "outdoors." Students tell me they enjoy the camaraderie of the camp, the communal division of tasks, and the work in the quiet landscapes at the edge of the lake. Most volunteers, I observe, leave the camp with an enhanced knowledge of local ecology and birds and with new emotional and affective ties to the Kızılırmak Delta's wetland ecologies.

The ornithologists have set up around forty nets in the area. Starting before sunrise and ending at sunset, students take turns walking from net to net, gently disentangling captured birds and placing them in small cloth bags, which they then mark with the net number. Yağmur, or, depending on the week, one of her colleagues, sits at the field lab desk and examines and measures each bird. She then places a thin and light metal ring, marked with a unique code, around each bird's leg. One student takes notes of the measurements and other characteristics in a field book; the stunned birds are then released outside. During this process Yağmur asks her students to identify the birds, and they flip through ornithological guides, aimlessly, until Yağmur shows them how to undertake the task, pointing out the correct species name.

Until 2011, the research station was lodged in the fishing cooperative's building, farther along the gravel road from its current location. Now, after a day's work, the researchers still sometimes walk down the road to drink tea with the fishermen. Scientists and fishermen use the same nets: fishers deploy them to catch carp, mullet, zander, and crayfish in the lake, and ornithologists use slightly modified versions to capture and study wetland birds. While ways of knowing the wetland through bird research and through fishing are in many ways contrasting kinds of practices, the nets remind one where these activities overlap: both rely on a knowledge of place emergent through practice, scientific and local knowledge, and national regulation—and for the fishers, market prices of fish. Fish and birds also become "sentinel" animals, through which both scientists and fishers can detect and assess the ongoing degradation of wetland environments.

With over 350 recorded bird species and millions of migratory and sedentary birds, the delta's local designation as "Bird Paradise" (*Kuş Cenneti*) is unsurprising. This term is not exclusive to the Kızılırmak Delta but common to all conservation wetlands in Turkey, reflecting mid-twentieth-century notions linking the value of wetlands to the provision of habitats for waterbirds. The term's enduring popularity in the twenty-first century is to the chagrin of many local scientists, such as Yağmur, who would prefer wetland visitors consider and appreciate the delta's ecological interconnectedness and biodiversity.

Yağmur and her colleagues' work in the Kızılırmak Delta is of the sort that historian of science Robert Kohler has called "residential." These are field ecology research practices based on long-term residence in place, whereby scientists come to know the specificities of environments, their human and nonhuman occupants, and the relations of different coexisting species. These environmental encounters, he argues, form the practices, theories, and ethics

Figure 4. A goldcrest (Regulus regulus) awaiting release from the ornithological research station, 2014. Photo by the author.

of the field scientists.[45] Care practices motivated by notions of moral ecology are in turn productive of place-making activities, such as those evidenced in the ornithological field station. Yağmur and her colleagues are deeply involved in multiple wetland advocacy initiatives, such as drafting wetland management plans and reports, applying for international conservation statuses such as Ramsar and UNESCO sites, writing grant applications, and participating in local public meetings. They also leverage their scientific work and their affective commitments to care for the delta to enlist the delta in national and international conservation, but this outcome alone does not explain their everyday moral, affective, and practical commitment to the delta and their expressions of ecological care.

The pointed use of care here foregrounds vital cultural and affective commitments. Care practices are multiple and intersecting. Care for the birds, for ecology, for students, for landscapes, and for birders, colleagues, and friends overlap. But while care sustains relationships—such as those between birders and fishermen, and students and the migratory birds—others are destroyed. For example, the implementation of environmental conservation boundaries and regulations in the delta generated new conflicts over rural residents' livelihoods. Scientists' notions of "good" and "bad" infrastructure in the delta are based on their understanding of wetland ecology as centered on their concerns with biodiversity, birds, and the social lives of researchers. And these notions fail to account for the care practices of rural residents—for instance, care for market crops, for household economies, tied to care for kinship, community, and social mobility. But who, exactly, is invited to participate in care of the delta as a wetland, and on what terms? The next section addresses the connection of conservation imperatives to varied notions of rural farmers' participation in the production of the delta wetlands.

Idioms of Participation, Practices of Exclusion

"Without accounting for the farmers' livelihood, we are bound to lose their support," Mustafa told a group of city officials and university professors in January 2016, during a meeting to plan for a wetland conference in the Kızılırmak Delta.

"Farmers have always opposed our work," an official replied. "I remember, back in the old days when we were first working to establish the boundaries of the conservation areas, they used to threaten to shoot at our cars."

"Without their support, conservation initiatives will fail," Mustafa insisted.

Mustafa's warning invoked the category "local people" (*yerel halk*), identified through personal encounters with delta farmers, as well as through pastoral ideals rooted in nationalist understandings of the connection between land and identity. This category "local people" has become central to Turkish wetland advocates' ecological imaginaries of the Kızılırmak Delta. Proponents of conservation recount conflict, sometimes violent, with delta residents in the 1980s and 1990s, as they worked to define the boundaries of the conservation area. Yet recently, many told me, local farmers have become more open to conservation initiatives.

Residential scientists develop deep personal and emotional attachments to the landscapes and waterscapes of the lower delta. They cultivate these attachments outside their research work, organizing field visits with their families, for instance, or taking and sharing nature photographs. Many of them, like Mustafa, also work to sustain ongoing relationships and friendships with delta farmers. Participatory practices like these are central to the processes of knowledge formation that legitimize scientists to speak for the wetland, as they position themselves as authoritative intermediaries between farmers, Turkish state and scientific institutions, and scientific organizations such as Ramsar.

During the course of my research in the Kızılırmak, I frequently heard the phrase "the delta has many owners, but no one who takes responsibility for it" (*deltanın sahibi çok, sorumlusu yok*). The ubiquity of this phrase among farmers, environmental advocates, and scientists resulted from its ability to be interpreted in two contrasting ways, with each interpretation reflecting a different understanding of delta governance. For some, this expression conveyed the perceived need for a stronger state presence in overseeing and coordinating the work of different departments and associations in the delta. Yet others envisioned greater grassroots "civil-society" (*sivil toplum*) participation in decision-making processes. Farmers are at the center of agricultural development plans for the region, but their perspectives remain marginal to those of conservation scientists and wetland advocates, even as they figure in imagination of the wetland as a site of valuable ecological livelihood and traditional practices.

One elderly rice farmer recounted his grievances during a conversation with me and other farmers in his house in May 2015. "The state [*devlet*]," he said, "came here and told us, 'You will grow this and this crop, in this, and this way.' Now it comes in and tells us, 'This is a conservation area. You can't hunt. You can't grow rice here.' Soon they will prohibit grazing our water buffaloes. The state just comes in and tells us what to do."

But Mehmet and Emine, an elderly couple who hosted me in 2014, had a different perspective. Mehmet enjoyed recounting how "Ankara" (by which he meant an official working for a government-funded program for community tourism development in the delta) had inspected and certified their house as appropriate for hosting tourists. Another small-scale farmer, Meryem, asked me if I could help her apply for state support to start a cooperative for selling local products made by women. And Ibrahim, a wealthier rice and buffalo farmer, planned to take advantage of state support to expand his farm and buffalo dairy. Many residents continued to perceive environmental conservation as an extension of state authority or as merely the interests of urban elites, which would invariably result in further marginalization. But other rural residents, already reliant on the state for pensions, health care, disability support, and agricultural credit and subsidies, also had expectations that the state would protect and support them in benefiting from wetland conservation, as state institutions did in other contexts.

In September 2014, as the "anthropologist-in-residence" in the delta village of Doğanca, I was asked on short notice to invite "local farmers" to speak to students at the wetland school. I tried, and failed, to find women who were available to participate. Some were reluctant, others were busy selling produce at the weekly town market, and still others were preparing food for a funeral. But three male farmers, whom I knew more closely because I had stayed in their houses, immediately agreed to come to the meeting. One of them, Ibrahim, who was then hosting me on the family farm, drove with me from the village to the wetland visitors' center where he was to address the group of students and professors to talk about his rice-farming practices.

As Ibrahim and I approached the village center, scattered farms surrounded by golden-brown rice fields and pepper fields dotted in green and red gave way to denser settlement. Ibrahim drove past the school; the mosque; and the old municipality building, now empty; the abandoned gas station; and an aspirational bus stop built for a route that never existed. Two-story houses in unpainted concrete hid behind low walls and metal gates alongside the road. Dogs ran out to the road to bark at the passersby. Pungent smells entered through the open windows of the car: the sticky smell of manure, the bittersweet scent of silage, the warm aromas of hay, the pungent smell of gas. These were tempered with the residual fragrance of firewood and boiling milk and the thickness of fried vegetables and meat stews.

This was the main road connecting Bafra to Doğanca and subsequently to the wetland conservation area, where it became narrow and unpaved. Visitors' experiences of the lower delta are invariably tied to the materiality and spatiality of this one paved road. We drove past the last rice and corn fields, finally reaching the marshes. The road finally passed directly in front of the visitors' center. Many local residents referred to it as *kule*, "the tower," because of its tall wooden bird observation structure. In a cloud of dust, Ibrahim turned to park between an excavator machine and a school bus. The university students were sitting down, in a semicircle, below the observation tower.

"What is the relationship between rice fields and the wetland lakes?" a student asked Ibrahim and the other farmers.

Others joined in. "Are farmers experiencing problems resulting from the overuse of pesticides and fertilizers? How do they introduce new crops and technology? What are the benefits and the drawbacks of letting the water buffaloes graze in the delta? Would it be possible to grow organic rice and vegetables?"

Ibrahim and the other farmers replied with stories and examples from their own farms, prompting the students to ask new questions. Here is an example of participation at work, I thought to myself while jotting down a quick note on my notebook: "farmers are performing expertise."

Soon, however, the professors began to argue among themselves about the students' questions. One declared that the only possible future for the delta was in organic agriculture. Another replied that this course would be practically impossible; farmers should instead be educated in the proper use of pesticides and fertilizers. Ibrahim sat back and listened politely, occasionally nodding.

Later, we drove back together to Ibrahim's family farm. He enjoyed the students' questions, he told me. But, he added, university professors consistently failed to address villagers' actual problems. A new disease had been destroying his and others' rice crops, he noted, particularly in the less windy areas. The delta's soil and water, he believed, could no longer sustain the heavy use of fertilizers and herbicides. The university, he added emphatically, did not reach out to the farmers to address their real needs.

Figure 5. University students at the wetland summer school meeting with delta experts and farmers, 2014. Photo by the author.

In contrast to approaches in participatory conservation that emphasize the involvement of different stakeholders, the ways in which wetland advocates implicate farmers are themselves implicated in historical imaginations of the Turkish peasant—fundamentally attached to romanticized ideas of the rural village as the repository of Anatolian tradition. On another day of the wetland school, scientists invited a very elderly farmer—and one of the wealthiest landlords in the delta—to tell the students about village traditions and customs. Scientists, students, and environmental advocates saw farmers like him as traditional subjects, spiritually and religiously close to the land, but having lost their past holistic relationship with production as they embraced modern development and in need of being taught, once again, how to farm without poisoning land and water and depleting resources. These ideas build on a longer legacy of Turkish nationalist imaginaries positing farmers to be the spiritual center of the nation, even when politically and economically marginal.[46]

Since the delta settlements resulted from shifting nationalist politics and population resettlements and migration in the last century, homogeneous "local people" exist only when invoked as such, forming an imagined delta community.[47] Similarly, farmers talked about "the state," or "the university" in general, as abstract and essential categories.[48] Yet these abstractions were made more real, in a sense, by the rapid transformation of land, water, infrastructure, and ecology that state and scientific institutions had produced. At the same time, the actual ways in which encounters between scientists and farmers unfolded reflected less the abstract ideological notions than the idiosyncratic friendships and collaborations that have flourished through years of scientific research and advocacy work in the field and sometimes shared, but often divergent, visions for the delta's futures.

Knowing Delta Waters

In the Kızılırmak Delta, residential scientists' care for the wetland arises from and is practiced through their material and scientific engagements with the multiple kinds of water and infrastructure that make and unmake the delta's ecology and biodiversity. Their knowledge and material engagements with water and infrastructure shed light on a multiplicity of visions of the delta as a scientific laboratory and a moral ecology. However, farmers in Doğanca often experience irrigation water as scarce and expensive. One day in the summer of 2015, Ibrahim was teaching me how to count banknotes by holding them between my fingers, tabulating the amount needed to pay the energy bill for the water pumps used to irrigate rice. More money—a set amount per crop, field area, and type of canal—would go to the irrigation union.

As I counted, we chatted about water infrastructure. Ibrahim believed that the delta's irrigation project, which had been decades in the making, always lagged behind agricultural expansion. For wet rice to grow, accurately regulated water flows are crucial. Yet access to scarce irrigation water, and the securing of the right provision of water at the right time, is mediated by personal relationships with irrigation authorities. Scarcity for smallholder farmers, water buffaloes grazing in the wet meadows, waterbirds, and aquatic species comes into being as water is provided in abundance for water-thirsty market crops such as rice. Rice growers like Ibrahim are aware that this crop demands more water than the new irrigation system could provide and needs more pesticide, fertilizer, and herbicide than is good for the delta's soil and ecology.

At the wetland school of 2014, students walked alongside irrigation and drainage canals and tested water quality in the wetland lakes. We learned about a newly built canal to fill the Cernek Lake with "clean" river water—but which remained empty because, according to university scientists at the wetland school, different state agencies could not agree on the amount of water that was in fact needed to sustain a wetland ecology. The question was not simply about water quantity (and quality) but also about when water would flow into the lake and how the varied effects of "irrigating the wetland" would be monitored. Students were concerned about the effects of water scarcity on wetland ecologies rather than on agriculture, yet they also learned to see the two as connected and to care about the effects of agricultural water flows on wetland ecology.

One afternoon all the students were struggling to keep their eyes open in the dark classroom in the delta's visitors' center. We had spent the previous days in the muddy meadows, nets in hands, capturing and naming all the insects we could catch and then collecting and drying plant specimens for a delta herbarium. An irrigation expert was lecturing about the delta's irrigation infrastructures, with slides upon slides of maps glowing in the dark. At the back of the room, someone was softly snoring. The professor mentioned the newly built canal redirecting runoff that previously flowed to the wetland lakes, to the sea, using a clicker pointer to trace it on the slideshow map.

Suddenly, a master student raised her hand and exclaimed, "But this canal just takes the pollutants somewhere else, to the sea! We are not addressing the root of the problem here."

Others woke up from their torpor and joined in, describing the cans of pesticides and herbicides abandoned in the canals they had observed during their field activities. The students talked about the shallow lakes getting shallower and eutrophic and debated ultimate responsibility for the wetland ecologies' futures. It was a heated conversation. Students were practicing their newly forged sense of care for this delta that they now had come to know as an endangered ecology. This was a sense of relatedness to delta ecologies, an emotional and moral commitment to care and take care, through their declarative statements and through the work of writing a petition to state authorities with a list of demands for environmental conservation.

The farmers' perspective on the canal was more directly connected to its varied effects on their farming practices and agrarian household economies. This was further complicated, however, by the presence of multiple and overlapping projects of delta development. These visions, once materialized through concrete infrastructure, had contrasting effects. Despite having initially agreed to the opening of the new drainage canal, farmers in Doğanca realized, after the fact, that the drying lake and the expansion of rice fields also meant less pasture for their water buffaloes. The number of water buffaloes had rapidly increased in the past decade thanks to a new program of state subsidies. Small-scale farmers faced increasing costs for feeding their water buffaloes because they had to purchase hay and feed supplements beyond the corn they already grew in their small fields. The rapidly changing and expanding cultivated landscapes of the lower delta also meant that the older common pastures, where many water buffaloes are left free to graze the entire summer, were now encroached on by new rice and corn fields. And lower levels of lake water greatly reduced the area of mud where water buffaloes could cool off from the summer's heat.

In the summer of 2014, a group of delta farmers took advantage of a technical "failure" in one of the cameras propped on the roof of the visitors' center. The cameras were officially used to perform bird counts and observe wildlife. They also served to control illegal hunting and fishing and other potentially illegal activities occurring in the conservation area. One evening, as the inexplicably broken cameras stopped recording, someone drove an excavator machine to crack the drainage canal open, letting water flow into the drying marshes at the lake's edge. This allowed the village's water buffaloes to graze and cool in the mud. It was in this water, flowing from a drainage canal to the marshes through this secretive act of everyday resistance, that later Mustafa would lead us into *becoming* water buffaloes and feeling the wetland.

Weeks later, I was driving to the lake one evening with village friends—all summer returnees to the village, in their fifties and seventies—for an evening picnic of rakı, grilled fish, and melon with their friends, who were fishermen. We came across the director of the conservation area, together with a small group of conservation staff, just as he had discovered the newly opened stream of water. For many state officials in his position, the preservation of the national and natural value of the wetland and their authority vis-à-vis colleagues and local rural communities depends on maintaining top-down regulations and controlling illegal resource use. The director was pacing back and forth, gesticulating while talking to his staff. We stopped and rolled down the car window to learn what had occurred.

"This is a crime!" the director exclaimed, turning to me to explain what had happened, pointing at water flowing on the grass toward the lake marshes. "Someone knew that the camera was not working and that I was not in the office—who did this?" he asked us, almost screaming.

My companions all feigned ignorance. However, the canal breach was an open secret in the delta villages nearby, and many people implied that they knew who had orchestrated it, and for whose benefit, although nobody shared their suspicions with the director. The newly created mudland would certainly not last long, however; the autumn rains would soon come, and then most of the buffaloes would be taken inside the barns for the winter.

Conclusion

The Kızılırmak Delta was reshaped through state-led displacement and relocation of ethnic and religious minorities, programs of malarial control, drainage and reclamation, agricultural expansion, water infrastructure, and top-down conservation measures. Its coastal marshes, in the process, were transformed into productive Turkish agro-economies. In the wake of these environmental, infrastructural, and demographic transformations, Turkish scientists participated in the international production of wetland science and conservation policy. These shifts were concurrent with the appropriation of the international category "wetland" into Turkish national imaginaries, as well as to the material remakings of environments. The delta emerged as a valuable wetland ecology, suitable for conservation, at the very peak of agricultural expansion in the delta.

While Turkey is often imagined as merely responsive to global currents of environmental conservation, anthropological analysis and histories of environments demonstrate the specificity of Turkish experience. To understand what it means to live in, care for, and contest ecological relations or moral ecologies, I analyze environmental transformations and histories of science alongside everyday practices of knowledge, care, and work. In places like the Kızılırmak Delta, Turkish scientists have formed varied senses of care and commitment to wetlands as valuable, fragile places. Their care practices are varied and multiple: care for birds, for teaching, for friendship, for leisure time, for nonhuman forms of life, for differentially imagined futures, for the past, for knowledge and science, for national development, for local governance. This work of care has also rendered the delta the subject of imaginaries of ecological value and virtue. These imaginaries are predicated on the continuing production of scientific knowledge and a hierarchy of expertise. The wetland school was one site of this ongoing production.

These senses of care contrast with the livable natures of rural residents, for whom the delta is also a site of production: of crops, animals, kinship ties, economy, identity, and social mobility. Yet wetland conservation in the current moment is actuated through situated encounters and collaborations between delta residents and scientists—invoking each other as stewards of each other's projects. For both groups, the delta is a place of care and work. In this sense, I approach wetlands as moral ecologies: the language of wetland conservation is not to be taken as a given but as a reflection of social and cultural positionings, contingent relations, and practices.

Foregrounding competing productions of place brings to the fore the political stakes of ecology. The poignant social and cultural meanings that wetland categories have acquired for different social groups in contemporary Turkey are rooted in the valuation of ecology. At a historical moment of increasing authoritarian rule and repression, wetland conservation provides a venue in which residents of Turkey advance the values of community formation, democratic scientific processes, and hope for shared futures.

5   Emergent Wetland Animals

Animal Conspiracies

One early afternoon in August 2014, I was resting in the shade of the front porch after hoeing in Efe and Fatma's small house garden, petting an orange dog everyone in the neighborhood called Tarçın, "Cinnamon." The house was in a neighborhood of Doğanca, a sprawling village in the Kızılırmak Delta near the wetland lakes. Efe's younger brother, Arda, and Arda's wife, Melek, invited me to join them as they went to visit their friend Kamil at his farm. The two siblings and their families were spending the summer in an apartment that Arda and his siblings had been building on the land where their parents' old house once stood: a simple wooden cabin topped with a thatched roof. In the new, two-story concrete apartment, we slept on the floor in small, unfurnished rooms without doors—the building was still under construction. I enjoyed the long afternoons spent at home with "Auntie" Fatma and several neighborhood women; drinking Turkish coffee; eating sunflower seeds, fruit, and cookies; and chatting about this and that while Fatma stitched small red, green, and gold decorative beads onto plastic slippers.

Doğanca farmers called this neighborhood Çorak, meaning, literally, "brackish, uncultivable, or barren." This denomination reflected the material characteristics of the wetland as its new settlers had encountered it in the 1950s, when it had become their new home. In this ecology of marshes, reeds, wet meadows, lagoons, and swamp forests many animals thrived: frogs, leeches, fish, water buffaloes, and over 350 species of birds. Before then, the delta had also been a cosmopolitan rural locale of villages connected to the commercial town of Bafra. But the Greek and Armenian neighborhoods, villages, and churches in Bafra, including two large Turkish-speaking Greek villages in the lower delta—only five kilometers from Çorak—had been depopulated and razed to the ground between 1916 and 1923, the surviving population sent on long death marches and then deported to Greece as refugees.[1]

For the Turkish farmers who subsequently migrated to the delta, and for the Balkan Muslims who were resettled there, life had not been easy. Arda's father and his siblings migrated in the early 1960s from a village near Aybastı in the mountains farther east on the Black Sea coast to what would later be called Doğanca. Their parents' first yield of wheat and corn in the delta had been bountiful, Arda told me, reporting on a much-heard family story, but then the land proved to be hardly cultivable. As a child, Arda recalled, he tended to the family's few water buffaloes, occasionally climbing on them, and was always caked in mud. But their home flooded often, and the family would walk over to the neighbors' house, which stayed dry. Soon Arda's father abandoned his attempts to cultivate cash-crop vegetables and grains and ran a teahouse in the village instead.

Arda and his siblings all left the delta to find more stable state employment in Turkey's largest cities. During my fieldwork in the delta, Arda and Melek's daughters were enrolled in two of Turkey's top universities, in Istanbul and in Ankara. Arda, now retired from his career in the army, was completing an online degree in sociology. But many of Arda's extended kin continued to live in Çorak year-round. Before the municipality had closed down in 2013, some had worked there. Others occasionally found employment in the cafeteria at the wetland visitors' center. Older women and men tended to the pepper and cabbage fields and to small herds of water buffaloes, cows, and the few remaining sheep. Young people aspired to move to urban centers. A branch of the family had migrated to Germany and returned occasionally during the summer holidays.

On our way to Kamil's house, we waved from the car at an elderly neighbor who was grazing her water buffaloes by the roadside and at another relative as she was heading back to work in her pepper fields. In recent decades, farmers had explained to me, cash-crop vegetable and rice fields had rapidly encroached on the village's common pastures. Many farmers were no longer able to easily take their buffaloes to graze in the wet meadows, a few kilometers away. With agricultural expansion, the conservation wetland had become an important resource for grazing livestock.

We passed through lush green corn and rice fields, driving on the gravel road alongside the drainage canal. I gazed along the canal's murky and green waters, observing it as might an ornithologist or birder, a habit I had picked up while participating in ornithological research and wetland education initiatives in the delta. I spotted a purple heron (*erguvani balıkçıl; Ardea purpurea*) and a little egret (*küçük akbalıkçıl; Egretta garzetta*). I was particularly fond of the little egret, one of the first Turkish birds I had learned to recognize. "The great white egret's beak is yellow, but the little egret's is black," a biologist working for an environmental NGO had taught me during a wetland bird count in the Aegean region the previous winter.

We continued past the old Bulgarian neighborhood, now almost abandoned, and a large farm estate. There we turned right into a neighborhood named after the Black Sea mountain village from where its current inhabitants had migrated in the late 1950s. On the unpaved road that ran through gray concrete houses and barns, two young boys and a girl, wearing sweatpants, T-shirts, and flip-flops, were all riding a squeaky red bicycle together, swerving across the road. The older boy was standing up on the pedals, as he could not yet reach the seat; another was sitting sideways on the frame, gripping the handlebar, and the girl balanced on the back rack. Alongside the road, women were at work in the vegetable gardens outside their houses. Water buffaloes and horses grazed inside a large fenced meadow on the other side of the road. The road ended at a small farm near a small forest of elm (*karaağaç*) and oak (*meşe ağacı*) trees. Fields planted with corn and rice surrounded the house and the forest.

We had reached the house of Kamil, a white-bearded man in his eighties. Kamil's son invited us to take a seat in the shade around a white plastic table. A horse was tied to a tree nearby. Kamil's daughters-in-law were moving about the barn, preparing to feed and milk their water buffaloes, and we were not introduced to them. Kamil's grandchildren, who had been biking together on the road, parked the bicycle against the wall of the house and brought us a large oval plate with white melon cubes and bright red watermelon, small forks, one water glass to share, and a plastic water jug.

Our conversation that afternoon centered on the changing population of wild animals in the delta. Kamil's son showed us blurry pictures of a jackal (*çakal*) he had taken with his phone. Jackals, foxes, and wild boars, once commonly encountered by farmers living in the lower delta, had disappeared in recent years, he explained. This might have happened after farmers cut down the swamp forests and transformed them into more fields and pastures, Kamil suggested. But then the jackals returned. Maybe the National Parks staff had reintroduced them, perhaps to counter the rapidly growing rabbit population, Kamil's son speculated.

A species of carp that recently started proliferating in the lower delta's lakes might have been willingly introduced by state officials, Arda added. The name of the fish, *Israil sazani,* meaning Israeli carp (but called Persian carp in English), helped fuel a wider conspiracy about foreign powers' threats. Kamil's son also mentioned the possible introduction of a characteristic bird, the purple swamphen (*saz horozu*) from another wetland conservation area in Turkey. Kamil disagreed: he had always seen the purple swamphens since he had migrated to the delta as a teenager, though he admitted that the bird had become more numerous, or perhaps more visible, in the previous two decades.

In contrast to the popular and widespread conspiracies about foreign countries stealing Turkish water, soil, seeds, and butterflies, delta farmers more often wondered whether rapid changes in wildlife populations resulted from the secretive actions of known political actors, state agencies or, alternatively, environmentalists (*çevreciler*).[2] Heightened preoccupations with wildlife invasions often reflect communities' worries about people taking advantage of resources at times of rapid economic and political change.[3] In the Kızılırmak Delta, the return of old species and the arrival of new ones added complexity to the already fraught process of deeming certain species as local (*yerel*), exotic (*egzotik*), foreign (*yabancı*), or invasive (*istilacı*). Water buffaloes themselves, now considered by wetland advocates to be one of the characterizing *natural* species of the delta, had almost disappeared and then recently returned, bolstered by a program of state subsidies and NGO projects funded by international donors.

Animals are political and social subjects and participants in human societies.[4] However, for Kamil and his family, animal taxonomies and agencies were not fixed entities but highly dynamic, entangled with and inseparable from human practices and politics. If humans are an "interspecies collaborative project," as anthropologist Deborah Bird-Rose and others have argued,[5] I follow anthropologist of science Stefan Helmreich in highlighting "symbiopolitics"—the politics of relations between living creatures.[6] My conversation with Kamil and his family about water buffalos, jackals, and birds was in fact a conversation about power, uncertainty, and environmental politics. This was an everyday politics extended to nonhuman animals but not fully detached from human politics. For delta farmers, water buffaloes, jackals, carp, and other animals were at once political subjects, agents of ecological change and place making, and companions of everyday life and work in the delta.

Animal studies scholars have in recent years moved away from symbolic approaches that considered animals as mirrors of human cultural concerns and social organization.[7] They have instead focused on multispecies practices, especially those concerning work,[8] forming kinship relatedness,[9] and probing ontological questions.[10] Yet it is still important to attend to how people make sense of varied animal practices,[11] in other words, how they engage with and theorize animals' agentive materiality and the ways in which this results in making animals into scientific and political subjects. Partly, I am foregrounding human protagonists here rather than shifting the focus onto animal perspectives and agencies. But animal practices are not reflective of human concerns. Rather, they conjure processes of environmental change and place making through embodied relationships between specific groups of people and specific animals—as individuals and as collectives.

The politics and poetics of wetland making, then, is necessarily entangled with animal politics. Environmental conservation is predicated on intersecting temporalities: connecting imaginations of past ecologies, whether of lost biodiversity or of environmental degradation, to desirable futures, and, conversely, drawing on calculations of future risk and loss to redesigning environments and policy. Creating a conservation area in an agricultural delta like the Kızılırmak involved deliberations over who had the "right" to live and work in the wetland, as well as over who and what are fit to represent the wetland politically and symbolically. In this context, the valuation and governance of animals are not only entangled with but are intrinsic to human politics.

Making Delta Animals

The simultaneous transformation of the Kızılırmak Delta's swamps and marshes into agricultural fields and into conservation wetlands affected the entangled lives of humans and other animals. Consider, for example, the effects of early twentieth-century marsh drainage on nomadic herders, migratory birds, and aquatic plants and the new relationships between farmers, plants, and domestic animals produced by cash-crop agriculture, alongside the displacement of forests, meadows, and wild animals to make place for chemically saturated monocultures of corn, rice, peppers, tomatoes, cabbage, leeks, tobacco, and sugar beets. Inside the wetland conservation areas, fishing regulations, hunting restrictions, and international markets for reed harvested in the wetland lakes have contributed to changing the delta's aquatic and terrestrial ecologies. For environmental managers and university scientists, building and maintaining the wetland as a conservation area also involve the daily work of monitoring, listing, counting species and individuals, evaluating their reproductive capacity, and addressing the removal of species deemed invasive or nonnative.

Drawing playfully from the work of cultural historian Raymond Williams,[12] I see animal practices—and the stories that farmers and scientists tell about them—as *disrupting* dominant ideologies of wetlands. Dominant wetland ideologies are the managerial and infrastructural work that state and provincial authorities undertook in the delta, typically following the Ramsar wetland playbooks and national wetland regulation. Dominant wetlands are crafted in the material remaking of water flows and ecological relations and become inscribed on the bodies of nonhuman animals. Dominant wetlands grapple with *residual* relationships: those left behind in changing agrarian ecologies or those elided through ethnonationalist transformations of human populations and their environments. Williams's notion of the residual is about the lingering presence of past relations as well as fantasies about that past. Deliberations over livelihood and belonging in the wetland generate new concepts and practices of wetlands' moral ecologies, what Williams would call *emergent*.

Like their colonial predecessors, postcolonial projects of environmental conservation entailed the displacement and dispossession of local and indigenous populations and, as a result, were invariably met with local resistance.[13] In the Kızılırmak Delta, contestations between different social groups, stemming from diverging understandings of the wetlands' ecological, economic, and cultural value, enrolled the livelihoods of other animals alongside humans in overlapping contexts. Farmers, scientists, residents, developers, state officials, and wetland advocates proclaimed contrasting understandings of wetland livelihoods in light of the transformation of the lower Kızılırmak Delta into an environmental conservation area.

In their conversations about delta animals as well as through their embodied participation in material practices of animal livelihoods, delta farmers, scientists, and activists claimed and debated desirable forms of wetland live-lihoods and politics. Claims about the moral ecology of the wetland were scripted through invocations of animal lives, sometimes as proxies for envi-ronmental politics that has become precluded in the intensification of author-itarian politics in contemporary Turkey. Nonhuman animals are central to both acts of resistance and compliance with regimes of environmental gover-nance in the delta's agricultural fields and wetlands. The rest of this chapter concerns the lively politics of wetland animals in the Kızılırmak Delta, focus-ing on water buffaloes, purple swamphens, and little egrets. First, I attend to the everyday work of the delta's water buffaloes, as these domestic animals conjured, for the delta's bureaucrats and farmers, contrasting visions of eco-logical and economic futures in the delta. Subsequently, I describe the making of the purple swamphen into an avian symbol for civil-society interest in the wetland conservation. Finally, I look at the issues of land access and marginal-ization as I examine why a colony of little egrets was accused of deforestation in the swamp.

Buffalo Biopolitics

One morning in August 2018, Ibrahim, a relatively wealthy delta farmer who had hosted me at his family farm for much of my fieldwork research, asked my husband, Ben, and me to accompany him to a large private hospital in Samsun for a medical checkup. There were two roads one could take to drive from the farm to Samsun. One, a forty-minute journey, followed the main village road toward the wetland conservation headquarters; the road then skirted the delta lakes and sand dunes, passed a lighthouse and a swamp forest, and joined the highway at Atakum (a municipality in Samsun Province). This road had been a gravel track during the duration of my long-term fieldwork. Recently, how-ever, it had been paved with asphalt. The other option took more than twice as long as the more scenic coastal alternative, and it involved driving to the town of Bafra and then getting on the highway to Samsun.

Ibrahim decided to take the shortest, fastest road. But things were compli-cated. Just a month before, as part of a new series of restrictions that govern-mental authorities justified by the UNESCO heritage site accession process, the coastal road had been closed to all automotive traffic inside the Ramsar conservation area. The stated aim was to protect the delta's wildlife from fast-speeding cars, pollution, and increasing streams of visitors. The road closure, environmental managers envisioned, would attract mindful ecotourists exploring the delta on bicycles, foot, or horseback. According to the new regulation, park officials and state officials could still operate their motor vehicles within the conservation area, to the chagrin of local rural residents who were precluded from doing the same. A "green" bus would shuttle visitors from the conservation area's boundary to the bird observation tower. However, rural residents could no longer use the road with their cars, tractors, or motorcycles—unless their water buffalo herds were grazing in the conservation area.

Ibrahim called up a couple of friends in the village to ask for confirmation that the road had indeed been closed to all residents. He decided to give it a try anyway. He probably knew the soldier posted at the gate, he reflected out loud after getting in the car. We were promptly stopped at the plastic road barrier, which had been equipped with a small control booth. Confidently, Ibrahim introduced himself to the young soldier as a "local farmer" and declared he was just going to check on his water buffaloes, which were grazing in the marshes beyond the road barrier. He waved a small, tattered square piece of paper as proof of his membership in the local buffalo breeders' association. The soldier, glancing at Ibrahim's document and ignoring my husband and me, immediately waved him through. At the next checkpoint, on the other side of the wetland conservation zone, Ibrahim explained to another young guard that he had just been checking on his water buffaloes grazing by the lakes by the side of the road. Again, the young guard waved him through.

This was a convenient expedient to negotiating the new road closure, but Ibrahim did go the delta's pastures almost daily to check on the family's buffalo herds. Rather than the flashy white car Ibrahim was driving on his way to the hospital that day, he and his relatives drove small scooters, rode on horseback, or piloted small dinghies in the water. The guards ignored the spotless car, Ibrahim's formal slacks and white shirt, and the presence of two foreign guests in the car, one in comically large sunglasses, jeans, and a short-sleeved blouse with a flower print and holding two large containers of buffalo cream and eggs on her knees (a gift for the doctor from Ibrahim's wife) and the other wearing a freshly laundered blue shirt. Neither of us looked like shepherds about to step into the marshes to herd their water buffaloes to a new grazing location. In this way, Ibrahim and the guards performed bureaucratic compliance with the new wetland management regulation while actually defying the rule in practice.

Since the delta became a Ramsar site in 1998, rural delta residents had worried that authorities would eventually prohibit their water buffaloes, horses, and sheep from grazing inside the conservation area, as they had done in other protected wetlands in Turkey. They were particularly concerned with the draconian grazing restrictions implemented in the Gediz Delta conservation area in Izmir. The transformation of this agrarian delta into a site of conservation had indeed entailed increasing restrictions on access and land use. Unbeknown to the delta farmers, in the summer of 2018 conservation authorities were preparing to prohibit people, including residents, to access the coastal dunes and beaches—a decision justified as part of the UNESCO accession process. Despite farmers' worries, however, the buffaloes were not displaced from the conservation area: on the contrary, they became central agents in state officials and university scientists' visions of agro-economic futures and biodiversity.

Domestic water buffaloes wintered on the delta's family farms. In the summer, most of the herds were released to the wet meadows and wetland lakes, where they grazed freely until the fall. But the population of water buffaloes in the Kızılırmak Delta had decreased dramatically, from about thirty thousand in the 1960s to two thousand in 2005.[14] This resulted from several interacting processes. In the 1960s, the Turkish government reduced its support for livestock breeding and new grazing restrictions to prevent land degradation and erosion. In the 1950s and 1960s, many farmers and shepherds migrated to Turkey's cities. Agricultural capital moved away from livestock herding to intensive cash crops for national and international markets. The introduction of tractors in the delta, and elsewhere in Turkey, starting from the 1950s had reduced the need for animal power in the fields: the labor of buffaloes (and horses) was replaced by fossil fuels.[15] In the delta, the expansion of rice and other cash-crop vegetables from the 1980s also reduced the collective pastures. The wetland conservation area then became an important resource for farmers to graze their sheep and water buffaloes.

In 2008, concerted NGO and governmental efforts brought the buffaloes back. A Turkish environmental NGO obtained United Nations (UN) funding for a project to rehabilitate wetland ecologies through water buffalo grazing "work" in the marshes—their grazing, fertilizing, and opening deep pools of mud. The NGO scientists argued that biodiverse wetland ecologies in the delta had been coproduced through long-term buffalo presence in the lower delta. The return of the buffaloes would also boost local economic livelihoods. The project report described water buffaloes in the delta as the largest "naturally" occurring water buffalo population in Turkey[16]—a perspective also espoused by veterinary scientists.[17] Project scientists strategically naturalized domestic water buffaloes, without compromising their domesticity, to support the argument that water buffaloes helped maintain wetland biodiversity. Thanks to the program of state subsidies that ensued, and the founding of a Buffalo Breeders Cooperative, farmers acquired new buffaloes. The delta's buffalo population grew from two thousand to over eight thousand in just five years, and buffaloes became once again central participants in rural wetland livelihoods and economies.

"I lost my mobile phone while I was taking my water buffaloes out in the delta for the summer," Mehmet, a retired construction worker and smallholder farmer, told me one morning early in May 2015. At the time, I lived in his and his wife, Emine's, modest farmhouse. I sensed from his expression that he was going to tell me an amusing story, and I waited for the punchline.

"Weeks later," Mehmet continued, "I was checking on my water buffalo herd in the *dil* (meaning literally "tongue," referring to a coastal area), and I saw something shine on the ground. I got off my tractor and took a close look. It was my phone! A buffalo had stepped on it and buried it in the mud. So I extracted it from the mud, cleaned it up, and dried it, and, look, it's still working!" Mehmet flipped his phone open and played a sound recording of his youngest grandchild.

It was early summer, and half of Mehmet and Emine's herd was already grazing in the wet meadows and lakes of the lower delta. A few times a week, Mehmet drove his tractor to check on the whereabouts of the buffaloes, redirecting the herd to a different area of the wetland if he thought they needed fresher pastures and regrouping the herd if it had dispersed. Other farmers navigated the marshy delta with scooters, bicycles, and small boats. Horses were the most effective way to travel the shallow lakes and marshes, but only the wealthier farmers still kept them.

After cleaning the barn and feeding and milking the water buffaloes, Mehmet, Emine, their daughter, and I sat down for a breakfast of yogurt, cheese, and meatballs—the milk and meat were from their water buffaloes. This was accompanied with sour cherry jam from the tree near the barn and cucumbers and tomatoes grown from seeds passed down from Mehmet's grandmother. They fertilized their house garden with buffalo manure, but Mehmet boosted his five *dönüm* of hybrid corn with chemical fertilizers and

Figure 6. Water buffaloes wallowing in the marshes, 2015. Photo by Gazanfer Demirer.

herbicides he purchased from the agricultural chemist in town. Emine put a large pot of milk on the boil on a wood stove in the courtyard. This was the stove she also used to bake large loaves of bread. A small refrigerated van stopped at the gate to pick up the bucket of cow milk Emine had left outside. The driver filled up Mehmet's milk production booklet, marking the day's delivery.

A few minutes later, Ahmed, a small business owner from Bafra, drove to the farm to pick up the six vats of water buffalo yogurt and two large tubs of water buffalo cheese he had ordered from the family. Ahmed talked with Mehmet about selling to higher-paying customers in the city of Samsun. He could put them in touch with wealthy customers in Samsun who would buy their buffalo dairy and also pay very well for it, he suggested. "No way. We just can't produce more. This is already too much work," Emine said. "I am tired," she complained. She was standing by the kitchen sink, gingerly rubbing salt on a fresh slab of cheese. "My hands hurt; my head hurts," she said. Twice a day, Emine milked all of the water buffaloes by hand. Emine's hands were becoming arthritic, and she also suffered from frequent headaches. "We are going to get rid of these water buffaloes next year," Mehmet added, repeating a common refrain. Their youngest daughter, who was my age and unmarried, worked alongside her parents on the farm. She struggled with a health condition and needed frequent medical treatment. Mehmet and his family hoped to sell the herd off and convert their house into a small family-run hostel for researchers, like me. The work of maintaining the herd had become too tiring and unforgiving for the aging couple and their daughter. But each buffalo had her or his own name, and Emine and Mehmet attributed the buffaloes with individual preferences for particular kinds of work, food, and touch. It was hard to let them go.

Water buffaloes and their human companions were also collaborators in wider biopolitical projects and labor economies, which anthropologists Heather Paxson, writing about artisanal cheesemakers in the United States, has described as "ecologies of production."[18] Twice a week, on market days, Mehmet and Emine woke up at two o'clock in the morning to tend to the buffaloes and cows, while their daughter, Gülbahar, cleaned the house and prepared breakfast. At six o'clock, they drove to the villagers' market in Bafra. Emine sold milk, cream, butter, and cheese in recycled plastic vats and bottles to her regular customers. In the meantime, Mehmet drove to make deliveries to clients at the veterinary's office, the forestry bureau, and an agricultural machineries store. Mehmet kept the cash earned from the sale. He was a strict manager of the household's modest finances. The income from the buffaloes' milk and meat supplemented his retirement payments and some disability support the couple received for their daughter. They had been able to build a new barn and purchase a tractor and, more recently, a car.

Like those of many farmers, Mehmet and Emine's new moral ecologies and moral economies as buffalo herders and dairy producers fit the provincial agriculture department's vision for increasing the water buffalo population, which involved an imagination of flourishing production of organic milk and meat in the delta, while also maintaining wetlands, as the buffaloes grazed in the conservation area. It became clear in my conversations with the buffalo cooperative's engineers, however, that the future of buffalo farming—as they envisioned it—lay in large industrial-scale factory farms, where the buffaloes would be indoor year-round, no longer spending the summer months in the conservation area's wet meadows and lakes. This vision excluded the majority of delta farmers, who were small-scale producers, like Emine and Mehmet, relying on the buffaloes to supplement seasonal working-class incomes, retirement payments, and cash-crop production.

The water buffalo revitalization project and subsidies had increased the number of buffaloes in the region. Yet it had not produced the coherent buffalo biopolitics that bureaucrats and agricultural engineers had hoped for. Farmers complained about the leadership of the breeders' association. The head of the association, some farmers speculated, had purchased a milk tank and a refrigerated van largely for his own use. Nobody, however, dared complain directly to the association or elect a different president, one of the farmers told me—the association still allowed access to precious resources and status, and farmers could not risk being left out. Buffalo biopolitics, then, did not instill bureaucratic compliance into the new buffalo economies. Most delta farmers did not keep consistent and accurate records of the daily production of milk in the booklets distributed by the breeders' association—much to the engineers' chagrin. Many farmers also shunned automatic milking pumps, attributing buffaloes' preference to milking by hand. Rather than sell their milk to dairy operations for national markets, most farmers sold their buffalo cheeses and creams to their personal acquaintances in nearby towns.

As buffaloes became biopolitical subjects, farmers could leverage them to advance everyday environmental politics in the delta. By expressing concerns with the possibility of future grazing restrictions in the conservation area, farmers called attention to the broader issue of their long-standing ongoing political marginalization from environmental governance in the delta—without having to expose themselves politically by making a direct claim against governmental institutions. Similarly, farmers like Ibrahim could leverage the necessity of their buffalo-herding work to bypass new access restrictions in the delta. Rather than a return to a putatively traditional buffalo economy, the modest sale of dairy products from the new herds generated small additional incomes, which supplemented pension payments, state employment salaries, cash-crop sales, seasonal wage-labor work, and urban remittances. Most important, perhaps, the growing number of water buffaloes meant that rural delta families now consumed their own buffalo cream, cheese, and meat almost daily, and new generations learned the art and science of home dairy production.

Delta farmers and scientists advanced varied, and contrasting, assessments of the environmental and economic changes animals induce in the areas they inhabit, through their livelihood practices. In this context, animal politics is one of belonging, whereby animals can become biopolitical subjects, unwanted others, or representative icons of biodiversity. These are not simple questionings of native-ness and invasiveness,[19] or questions of economic productivity and environmental sustainability, but moral assessments of the animal's unstable and shifting statuses in the context of agrarian change.

The Elusive Purple Swamphen

In August 2012, I met the director of the Kızılırmak Delta wetland conservation area—a civil servant at the provincial National Parks tasked with overseeing implementation of conservation governance and hunting control in the delta. The director was also charged with welcoming and providing assistance to visitors and researchers. On our first encounter, enabled by a well-known local university ornithologist, the director offered to take me on a daylong tour of the lower delta with his park rangers. After driving for several hours through the conservation area's swamps, beaches, and lagoons, the director stopped the car on an earth dike separating rice fields and wet meadows. We stepped outside to look at grazing herds of cows and water buffaloes, surrounded by flocks of birds in the sky. "They are practicing their flight strategies before the migration," the director explained excitedly. Since he had been appointed in the delta's conservation area, he recounted, he had developed a genuine interest in learning about ornithology and wetland ecology.

Suddenly, we heard several gunshots nearby. Soon an argument ensued between the director and a man, smoking rifle in hand, who suddenly stepped out of the rice fields onto the dike. The farmer explained angrily that he was chasing a purple swamphen from his field. The birds had been damaging his rice crops, he exclaimed. As he raised his voice, a gray heron took flight. Leveraging my presence, the director responded, speaking loud enough so that I could hear him: "These are internationally protected wetlands, and this is a bird of international importance [ulusalarası önemi]. These are protected birds [koruma altında]. You can't hurt them!" The farmer, unfazed, insisted that rice-loving birds like the purple swamphen had been eating his crop. The director remarked that the land on which the farmer had planted rice belonged to the State Treasury: the farmer, he reminded him, had leased it from the state. After this performance of political jurisdiction over the birds and the rice, the conversation between the two men became softer and quieter: I was no longer an audience for it. As the farmer insisted on his right to shoot the bird, the director added, more gently, that he could come to his office and talk about it over tea and apply to get monetary compensation. He added that he, too, was born in a village nearby. The two men then moved to talking about common relatives and acquaintances, and, shortly after, we drove on.

The purple swamphen, an elusive, rice-loving wetland bird, appeared on the radar of Turkish ornithologists in the Kızılırmak Delta in the 1990s, became an endemic and endangered species, and then became a symbol of civil-society organizing in the delta. The livelihood of the purple swamphen is vitally connected to the many Mediterranean and Black Sea wetlands that were dramatically drained over the course of the twentieth century. In the Kızılırmak Delta, the alleged return of the bird in the 1990s demonstrated at once the success of environmental conservation and its demise in the face of industrial agricultural expansion. As groups of farmers, scientists, and activists made meaning through the livelihood practices of the swamphen, they also articulated ongoing uncertainty about ecological change and human and nonhuman livelihoods in an agrarian wetland.

The swamphen's livelihood practices in the rice fields also made apparent the contested geography of land use at the edge of the conservation wetland. Land tenure and use in the lower delta are quite complicated—at least, for the visiting anthropologist. Property and use rights are not always formally registered as title deeds upon inheritance, my hosts in the delta explained. Rather, they are often agreed on among family members. These arrangements lead to occasional conflict. Despite the legal absence of categories of common property, residents follow customary norms for the communal pastures. Fishing, by contrast, is regulated by fishing cooperatives that legally operate in each body of water, with established fishing quotas, prices, and seasonal restrictions. Large sections of agricultural land reclaimed from marshes in the mid-twentieth century belong to the State Treasury and are leased to individual farmers for agricultural cultivation. They are irrigated and drained through infrastructure constructed by the DSI and managed by a local irrigation cooperative.[20]

Inside the conservation area, where buffaloes, sheep, and horses graze unrestricted, agriculture and hunting are prohibited. A system of yearly quotas and permits regulates commercial reed cutting. For farmers, the abstract governance structure of the wetland conservation area becomes tangible in this system of institutions and permits, expressed in everyday encounters with state officials tasked with implementation and enforcement—like the one I had just witnessed. These encounters are also mediated by class, kinship, and

Figure 7. Reed cutters at work in Cernek Lake, 2015. Photo by Gazanfer Demirer.

other social and power relations and axes of identity. In this case, my presence and the invocation of kinship ties allowed a rapid deescalation of the confrontation between the director and the farmer/hunter. I wondered afterward, when I got to know the farmer and his family, who never again mentioned this episode to me, if the whole scene had been a performance staged for my benefit—and, if so, what was I to learn?

The purple swamphen is an entry point into how delta residents, managers, scientists, and civil-society activists articulate claims over ecological and agro-economic change in the delta. This purple swamphen is widespread in Mediterranean wetlands, and other species of purple swamphen are found in Africa, the Americas, the Pacific, and Southeast Asia.[21] Although the swamphen lives in the wild and at the edges of agricultural fields, it also has long been domesticated. In ancient Egypt and Rome and in Byzantine times, people kept purple swamphens in their houses and gardens as a companion bird and traded it across the Mediterranean.[22] Today, the IUCN lists the Mediterranean purple swamphen, like the one that was escaping from the armed rice farmer in the delta, as a wild endangered species.[23]

Purple swamphens tend to be sedentary but can also fly long distances. They generally live in areas of shallow water, among reeds and grasses. They eat shoots, leaves, roots, stems, flowers, seeds, rice, grasses, sedges, water lilies, clovers, ferns, mollusks, leeches, small crustaceans, and insects.[24] For the environmental conservationists I talked to during my fieldwork in the Kızılırmak Delta, the purple swamphen added further national and international value to the delta's biodiversity. For the nature photographers who sought charismatic close-ups of delta birds on the weekends, the swamphen was an aesthetically pleasing bird to photograph in the wild, with its iridescent green and blue feathers and bright red beak. Rice farmers, like the families who hosted me in the delta during my research, generally considered it a benevolent bird that could become a pest during the rice season.

Rather than chase the birds out of the fields with a rifle, the wealthiest landowners more often would set up automatic propane cannons—these went off at regular intervals at the edge of their rice fields, keeping the birds away (and keeping me awake at night). Years after the encounter between the director and the farmers, as I got to know several delta hunters, I wondered whether the farmer had actually been chasing the swamphen out of the rice field or if the bird had just provided a convenient excuse to avoid a hefty fine for duck hunting. I would never find out. Either way, in talking about purple swamphen practices, the two men were articulating a tension between conservation imperatives in the delta and the simultaneous state-led vision of agricultural intensification.

One year after this first encounter, the purple swamphen became a symbol of environmentalist and civil-society interest in the conservation of the Kızılırmak Delta. In October 2013, the university hosted Turkey's third National Wetland Conference, organized in collaboration with the Ministry of Forestry and Water and its Wetlands Bureau. The organizing committee at the university commissioned a local graphic designer to draw a stylized purple swamphen. One of the conference organizers, a professor at the local university, later told me that they had chosen the bird because it straddled wetlands and rice fields alike and also because it had such a peculiar and attractive shape. This design became the logo printed on all conference posters, reports, publications, and promotional merchandise. In the spring of 2014, the same logo was used for the Nature School, a training camp for university students organized by university scientists and environmental educators with financial support from Turkey's Scientific Research Council (TÜBITAK).[25] The bird had been appropriated as a new symbol of civil-society involvement in study-ing and preserving the delta's biocultures.

The students appropriated the purple swamphen as a meaningful icon. In late September 2014, at the end of one session of the Nature School, the group of university students who had participated in the two-week-long edu-cation camp decided among themselves to offer a small gift to the organizing team of professors and educators, with which they would surprise them at the school's closing ceremony. Each student donated a few liras, and a twenty-year-old took the lead. Eren, born and raised in Bafra, was studying to take the national teachers' examinations. Since he also volunteered as a tutor at the municipal Popular Education Center, he contacted the center's glassmak-ing teacher and commissioned her to create small purple swamphen figurines made from recycled glass. At the Nature School's closing ceremony, the stu-dents proudly presented each of their teachers with a glass purple swamphen.

As the bird's rise to fame continued, its symbolic meaning became institu-tionally entrenched—at least, for a brief moment. In February 2016, Samsun Municipality called a press conference to announce that the purple swamphen had become the "official" symbol of the Kızılırmak Delta's wetlands. The delta, a municipal official declared, hosted the second-largest population of purple swamphens in Turkey—about fifteen hundred birds. The municipality pledged to help the population grow even more numerous. It had also begun featuring the swamphen in the city's promotional material advertising the delta. On a rainy evening in early January 2016, I turned to see an oversized purple swamphen towering over a bus stop near the Samsun airport, where I had just landed. The billboard advertised the varied fauna of the Kızılırmak Delta conservation area, eighty kilometers away. In the six months I had been away, the city trains had been covered in bird photographs advertising the delta's wetlands. Road signs started at the airport and continued until the conservation area.

I drove from Samsun to the village of Doğanca with a friend, a retired engineer and amateur nature photographer. After the city gave way to orchards, fields, and villages, signs with photographs of wetland birds con-tinued to mark the road to the wetland. Outside the village of Yörükler was a large billboard featuring galloping water buffaloes and horses, welcoming visitors to the delta. A gate of sorts had been erected across the gravel road, topped with plastic sculptures of swamphens, cranes, herons, and storks. Near the large wetland lake, yellow diamond signs warned of water buffaloes crossing. Nearby, herds of water buffaloes were muddling in the wetland lakes and grazing in the wet meadows. I looked for herds of buffaloes belonging to my hosts in Doğanca, squinting my eyes to try to see the family marks burned on their skin. Only birds, buffaloes, and horses featured in the conservation area's billboards and signs. The delta's rural residents, and their cows, sheep, dogs, ducks, geese, and chickens, and the delta's fish, amphibians, and insects were absent. I was not surprised by this omission. Twentieth-century imperatives of wetland conservation stemmed from the aviary concerns of birders and hunters. They incorporated national and colonial practices and imaginations of nature, race, and economy.[26] Charismatic animals, especially eye-catching birds and charming mammals, continue to be used as iconic symbols of nature conservation.[27]

The new signs were a stark visual intervention in the delta's varied agricultural, residential, forested, and wetland environments. By symbolically connecting the wetlands, a one-hour drive away, to Samsun's airport and city center, the signs also marked the province's administrative oversight of the lower delta's development as an area of tourism, research, and agricultural production. By expanding the symbolic presence of the Kızılırmak Delta conservation area to regional, national, and international nodes of transportation, they connected a local conservation site to national and international imperatives of wetland conservation. Yet what interested me was not the semiotics of governance but the everyday political lives of wetland animals in the marshes, meadows, farms, fields, and forests. Daily life in the Kızılırmak Delta for the farmers, wetland scientists, and conservation officials I worked with unfolded through entanglements with birds, fish, amphibians, bovines, horses, sheep, dogs, and so forth. These are not just symbolic or affective relationships but shared practices that sustain, and in turn are shaped by, working ecologies.

A IUCN report on endemic species, the delta management plan, and countless other scientific publications and newspaper coverage all described the purple swamphen as an endangered species (*nesli tukenen canlı*) native (*yerel*) to the delta. Yet many people suspected the bird was a recent newcomer. The bird moved between the statuses of native and a latecomer, pest and valuable symbol. Questionings of the bird's belonging, among farmers, scientists, and civil-society activists, also reflected broader concerns with ecological and agro-economic transformation in the delta and articulated uncertain moral ecologies.

Making and Unmaking the Purple Swamphen

The purple swamphen became a registered Kızılırmak delta bird-resident in the 1990s. The 1992 bird census, conducted by scientists at the Turkish foundation DHKD, the Dutch foundation Working Group on International Water and Waterfowl Research (WIWO), and the Ornithological Society of the Middle East, had reported no sightings of the purple swamphen.[28] The ornithological guide *Birds of Turkey* claimed that rural delta residents first reported spotting the purple swamphen in 1993.[29] In 1994, the Ornithological Society of the Middle East's newsletter *Sandgrouse* reported that delta hunters had shot three or four purple swamphens.[30] In an episode of the documentary *Wetlands Sources of Life*, broadcast on the Turkish national channel TRT1 in 2000, the director boasted that the documentary team had sighted the bird in the delta for the very first time.[31] By 2007, a local master's student had estimated the presence of thirteen hundred to fifteen hundred swamphens in the delta, including 250–500 breeding pairs.[32]

Many elderly farmers in the Kızılırmak Delta told me they had encountered the purple swamphen for decades. The expansion of rice had made the bird at once more visible and more numerous, a farmer and his brother explained to me one night, drinking tea and eating halva after dinner at home. However, on a different occasion, younger delta farmers told me that they thought the swamphen had only very recently arrived from the Göksu Delta in the south of Turkey. Perhaps, they alleged, the bird had been introduced by environmentalists or by the state. These discussions of the bird's supposed arrival and its livelihood practices conjured different social groups' preoccupations with the agro-economic changes of the delta and its concurrent transformation into a conservation area. In talking about the swamphen, farmers also reckoned with the political effects of state and nonstate institutions' claims to ecological stewardship in the delta.

The taxonomy of the purple swamphen itself had been the subject of scientific debate. At different times in the twentieth century, scientists had classified it into four, three, six, and seven species and into different subgroups.[33] Ornithologists working at the university in Samsun told me they subscribed to a taxonomical understanding that posited the existence of only one global species of *Porphyrio porphyria*, with thirteen subspecies—two of which, *caspius* and *seistanicus*, are found in Turkey.[34] I emailed Yağmur, the university ornithologist in Samsun to ask whether the swamphen had indeed been introduced from the Göksu Delta. She clarified that the Kızılırmak and the Göksu Deltas' purple swamphens in fact belonged to two different subgroups, *caspius* and *seistanicus*, respectively.

An ornithological study conducted by Turkish NGO and state scientists in 2008 in the Göksu Delta proposed two hypotheses for the growth of the swamphen population in the Kızılırmak Delta in the previous decade. One hypothesis posited that the swamphen had lived in the area for centuries, hardly ever seen, hiding in the thick reeds; another, that the swamphen had migrated in the previous two decades from the Caspian Sea through eastern Turkey. Once widespread in all European wetlands, swamphen populations declined as a result of the drastic drainage measures of the twentieth century, and the bird found refuge in the remaining wetlands. The bird's revival, then, according to ornithologists and conservationists, correlated with the "success" of wetland conservation policies.[35]

As historian Harriet Ritvo has shown in her work on Victorian-era naturalists, in debates about animal taxonomy species boundaries are unstable and shifting and often reflect human politics.[36] Debates over purple swamphen classification, its natural habitat, and its migratory history elided the long history of human relationships with the swamphen, including the bird's role as household companion, ritual animal, and trade object since at least 500 BCE.[37] The swamphen was made and remade, from lost species to elusive bird, icon of biodiversity conservation, and agricultural pest. This is similar to anthropologist Celia Lowe's account of how the Togean macaque in an Indonesian conservation area was made, unmade, and remade from an unnatural hybrid swarm to an endemic and charismatic species.[38] In the 1990s, the purple swamphen had become an elusive delta species, at the center of competing notions of what counts for an authoritative sighting: for example, birders' records, a rural hunter's account, or videographic evidence. By the 2000s, debate ensued about the taxonomic status of the purple swamphen and about whether it had been introduced or migrated to the delta.

Allegations about the introduction of the swamphen speak to a broader experience of the Turkish state presence in the lives of farmers and also to the demographic composition of the delta, a receiver of migrants from other rural areas of Turkey since the 1950s. After the city and the University of Samsun made the purple swamphen into a logo, and a symbol, of the delta's conservation area, the bird was stabilized into an endemic species. The appeal of the swamphen was partly aesthetic: nature lovers were drawn to its red beak, iridescent blue and green plumage, and thick red legs. Birders and photographers enjoyed observing the ways in which the purple swamphen dives in the shallow wetland lakes, nests in the thick reeds, and climbs on tall stems. Adding to its charisma was certainly the fact that the purple swamphen can be difficult to see and even more difficult to photograph.

This elusive quality had also to do with the delta's environmental change rather than merely with bird practices. Between 1992, the year of the first bird census, and 1998, when the delta was added to the Ramsar list, the delta's wetlands were reduced from fifty to twenty acres. The conversion of marshes, swamps, and wet meadows to rice and cash-crop fields continued steadily up to the time of my fieldwork. The purple swamphen, which usually nested and fed in thick reeds, found new sources of food in the sprawling rice fields. It became more visible (to humans). For farmers, then, the swamphen *became* a pest when they switched to rice farming. These conflicting interpretations of the bird status are best understood within changing agro-economies and landscapes, attending to changing animal practice in times of environmental uncertainty.

Contested Avian Symbols

Discussions of the status of the purple swamphen and its new denomination as a symbol of environmental conservation in the delta also reflected broader concerns about shifting environmental and political governance of the wetland. In November 2015, a large plastic sculpture of the swamphen and other delta birds was placed on a new ornamental gate, marking the entrance to the delta's conservation area across the newly asphalted road. Like the road repaving process, the sign itself had caused debate among wetland advocates. Partly, the debates were not about the birds themselves but about the sign's materialization of municipal boundaries in the conservation area. Initially, the city's employees had placed the gate farther away, near a lighthouse, well inside the conservation boundaries. Scientists and members of the Kızılırmak Delta Volunteers Platform—a small forum in Bafra and Samsun representing civil society's interest in the delta—pressured the city to move the sign back by a few hundred meters to mark the geographical boundary of the Ramsar area.

Not only was the position of the sign contested, but its content also raised questions about the ecological value of the delta and its ownership. The initial writing on the gate, designed by the director's office, read (in Turkish)

"Welcome to the Bird Paradise." It immediately generated criticism among university ornithologists and environmental advocates, as it conveyed the sense that the delta's value was merely as a habitat for birds. The wording "Bird Paradise," wetland advocates often told me, generated expectations in the weekend visitors from the nearby urban centers of Bafra and Samsun that they were to see caged bird displays. "We couldn't see any birds in the bird paradise," residents of Bafra or Samsun often told me in our conversations about the wetland conservation area. Instead, scientists and educators wanted to convey to the public, and to rural residents, an ecosystem understanding of the delta's value. The municipality, in turn, sought to emphasize the economic value of agriculture and promote, or, at least, advertise sustainable development.

Eventually, the signage was changed to a general term: "Welcome to Samsun's Kızılırmak Delta."[39] Even this geographical specification became the subject of debates. The municipality of Samsun is approximately fifty kilometers from the delta, though its province encompasses the delta. Should the name of the closest municipality be included? Or was it to be named after adjacent villages? I followed the debate's ramifications online.[40] In November 2015, an environmental NGO worker and conservation expert I had worked with posted on the Facebook page of the Kızılırmak Delta Volunteers Platform, questioning the word "Samsun" on the gate of the conservation area. "Has the Kızılırmak River secretly formed another delta in Samsun?" he joked. This joke played on the suggestion of a conspiracy involving Samsun Municipality secretly forming parallel deltas, echoing decades of "deep" and "parallel state" conspiracies in Turkish politics.[41] A state official in the conservation department replied that since there had been countless arguments about which of the delta municipalities could claim the delta as their own, designating it as Samsun would name the province that encompassed all other municipalities and districts and end the debate. His interlocutor responded that this denomination was a showcasing of power on the part of municipal and provincial authorities, who would now be "taking ownership over" the delta, excluding other constituencies (local residents, scientists, and environmentalists). An ornithologist chimed in to say she thought the delta should not be associated with any provincial or municipal boundary, and she was pleased that, at least, the wording "Bird Paradise" had been removed. The same ornithologist commented on the plastic sculptures of the purple swamphen and of a crane. The official symbol, according to the delta's management plan, is in fact the common crane (*Grus grus—turna* in Turkish) and not the swamphen, which, she explained, is the official symbol of the Göksu Delta. A bird photographer chimed in to joke, "At least they didn't put a penguin!" In subsequent years, the swamphen symbol disappeared from municipal and provincial material for the delta: the delta management union, founded as a partnership between municipalities and the Samsun Province, featured a crane as its official logo. However, local news reports continued to describe the swamphen as the "official" symbol of the delta.

Little Egrets and Disappearing Forests

Behind the Kaymak family's compound, near an apple and fig orchard that, according to the older Kaymak relatives, is planted over an old Greek Orthodox cemetery, is a small patch of swamp forest. I always went to walk in the forest whenever I needed some quiet time away from the busy dynamics of intergenerational family life on the farm. Stork nests are perched on the tallest trees, while little egrets (*küçük ak balıkçıl*) prefer the thick of the forest. The swamp extends until the main road and the rice irrigation canals, and members of the Kaymak family graze their cattle, sheep, and horses there in the spring. In the winter, the forest is immersed in water and deep mud.

One day in May 2015, three ornithologists I knew came to the farm to monitor the little egret nests in the forest. While the little egret is a commonly spotted along the irrigation and drainage canals and on the shores of the lakes in the Kızılırmak Delta, less known are the birds' nesting sites in the delta's few remaining patches of forest swamps. Ibrahim, one of my hosts at the farm, asked me to guide the ornithologists through the forest. I followed a narrow and muddy path marked by hoof- and footprints deep into the forest, looking for the ornithologists, who had gone ahead. Unable to find them, I retraced my steps to a parked car, where the ornithologists, in khaki and green outdoor clothes, were standing, quietly observing the stork nests and marking field notes in a notebook.

We walked together back into the forest, balancing our steps on grassy clumps, interspersing our footprints with other mammals' tracks in the mud. Suddenly, we heard gurgling calls and the rustling of leaves. Following muddy clearings in the vegetation, we reached a clearing covered in white droppings. Whiffs of a strong acidic scent reached my nose. We looked up. High in the

Figure 8. A field sketch of little egrets (Egretta garzetta), 2015. Drawing by the author.

trees was a small avian metropolis. Hundreds of little egrets were flying from tree to tree, calling loudly, perhaps disturbed by our unexpected arrival. The ornithologists observed the nests for a few minutes with their binoculars. They took note of the nest and bird counts in their notebooks, and, as quietly as we got there, we left.

Another close encounter with the little egret brought to the fore the entanglement of animal practices in contestations over environmental change and land use in the delta. On a scorching afternoon in June 2014, the director of the Samsun bureau of the Water and Forestry Department, a forest engineer, joined a group of Turkish university students, professors, and me at a session of the wetland school. We drove past the rice fields of Doğanca, parked the bus and cars near an empty barn, and continued on foot—balancing our steps on clots of soil and dry grass on liquid mud and fresh buffalo dung. We reached a large open area: hundreds of dry, leafless trees and countless dead tree stumps on the ground, all eaten by termites and decomposing. Little egrets began calling loudly as we approached.

We gathered around a large stump. The forester explained to the students that in recent years the colony of egrets had become more numerous and more concentrated. The birds' droppings, accumulating underneath the trees where they nested, had increased the salinity and acidity of the soil, killing the forest. The birds were responsible, he concluded, and must be relocated.

A professor of agricultural engineering scowled. Cutting the forester off, the professor launched into a different exposition. "The birds are innocent," he declared. Rice growers, he continued, had *illegally* extracted groundwater and drained the delta's wetlands, diverting water away from the forest.

"This is what happens with indiscriminate resources use in wetland areas," the professor exclaimed, pointing at the dying trees. "State authorities should implement environmental regulations more thoroughly," he added, glaring at the forester.

The forester rebutted, "The little egret theory of forest degradation has indeed been supported by French scientists." He had himself seen examples of this during an official study visit to a French wetland site in the Camargue at the Tour du Valat wetland research station.

"This is false and unscientific," the professor responded. Instead of worrying about little egrets, state authorities had better address the ongoing problem of wetland encroachment.

The forester and the professor continued arguing heatedly—none of us dared to intervene—until the forester, outraged, turned back and walked away to his car.

Back in the village, when I asked my hosts about the fate of the forest, I heard a third explanation. Two years earlier, a new forestry law allowed villagers who demonstrated previous use and customary rights to claim land that had lost its forest characteristics and buy it at a subsidized price from the State Treasury in installments. This legislation allowed informal rural occupants to obtain ownership deeds for their customary groves, orchards, and pastures. The law, however, also gave amnesty to thousands of houses built on forestry land on the Turkish coasts and in expanding cities.[42]

Despite the forester's understanding of the farmers living at the edge of the conservation area as a profit-driven environmental hazard, my hosts in the village made decisions predicated on their moral ecology and inflected by their own positioning in overlapping networks of responsibility, economic production, and care. This forest had become a place where impoverished local farmers rushed to get wood before others claimed it. And the rapid expansion of rice fields since the 1980s had decreased the pastureland for water buffaloes to graze, thereby turning this forest into a profitable pasture. With the rapid disappearance of other swamp forests, the little egrets had become more visible.

The little egrets, then, were cast as "matter out of place" in the particular encounter just described.[43] The dialogue was also a staging of ecological causality and a questioning of responsibility for environmental degradation. For delta ornithologists, the little egret story was a parable demonstrating the loss of bird habitat in encroached and rapidly disappearing wetlands—a similar narrative to that which motivated efforts to enact wetland preservation in the early and mid-twentieth century. For state officials, the death of the forest proved the culpability of the birds themselves, complicit in their own habitat's destruction. The egrets provided an intervention on the local scale, devoid of human politics, to address thorny questions of environmental conservation and degradation. For university scientists, state authorities were responsible, through their institutional neglect of the delta, for both small- and large-scale environmental degradation. By failing to enforce environmental protection regulations, they implicitly authorized forest and wetland loss. From the perspective of delta farmers, the birds, like the farmers themselves, were caught in a struggle for resources in a changing ecology of intensifying agrarian production.

While environmental conservationists ascribe moral value to Turkey's wetland ecosystems for the life-forms they support—animals, plants, microbes, and, more recently, people—life in the Kızılırmak Delta wetlands is shaped by the interactions of changing conservation policies with agro-economies, refracted in the everyday interactions among humans, animals, water, soil, and plants. Wetland animals such as water buffaloes, purple swamphens, and little egrets have become proxies for conversations and contestations about environmental change and politics of responsibility. These debates are not merely symbolic but also revolve around the material practices of individual animals and animal communities in specific places in the delta.

Conclusion

The establishment of environmental conservation areas has frequently resulted in the displacement and marginalization of local and indigenous populations; as a result, conservation projects were always met with local contestation and resistance.[44] Scientists and environmental advocates in the Kızılırmak Delta have enrolled delta animals—both wild and domestic, native and introduced, feathery, furry, and covered in scales—as living symbols of the ecological and social value of the wetland. Yet representations of wetland animals were contested in everyday conversations as well as in online forums, classrooms, conferences, and official meetings of wetland bureaucrats. These contestations centered on fundamental questions about the nature of the delta's lived ecologies, its residents' livelihoods, and possible futures: they were moral assessments of human and nonhuman entanglements in the agrarian wetland.

The shared lives of people and animals in the delta raise important questions about life, labor, and environmental politics in contemporary Turkey. Birds, fish, slugs, sheep, horses, and water buffaloes have become essential protagonists in Turkish people's imaginations of wetlands as beacons of biodiversity and conservation. Alongside this transformation into charismatic symbols of conservation stewardship, nonhuman animals have also been the subjects of debates over who belongs to certain ecologies and what their livelihoods should be.

Experiences of ecological life, agrarian work, and environmental gover-nance in the Kızılırmak Delta were scripted through invocations of bovine and avian lives, as lively proxies for profound questions of environmental politics. Nonhuman animals are central to both resistance to and compliance with environmental governance in the delta's agricultural fields and wetlands, both in everyday conversations about delta animals and in their embodied participation in material practices of animal livelihood and ecological work. These processes and relations constituted everyday environmental politics and fertile ground for lived and contested moral ecologies.

Conclusion

One day in early August 2018, Ibrahim and I went to check on the work of a crew of women he had hired to weed his rice fields. We were in the lower Kızılırmak Delta, where Ibrahim and his brothers' fields butted up against the large lakes just outside the boundary of the wetland conservation area. Driving us on his tractor through his neighborhood, Ibrahim glanced at his neighbor's paddy and noticed, from a localized discoloration in the plant, signs of blast—a fungal disease (*Pyricularia oryzae*) that Ibrahim referred to as *yanıklık hastalığı*, the "burning disease."

"He should really apply fungicide," Ibrahim told me, "but it will be very expensive." He paused for a few seconds. "In any case," he continued, "it's too late. He's going to lose half of his yield."

"Why?" I asked him.

"Rice is alive, like a person." Ibrahim likened the rice plant at this stage of its livelihood, carrying grains of rice on the panicle, to a "pregnant woman." Once rice gets to that degree of rot, Ibrahim explained, one would need to use a large amount of chemicals to hope to combat the fungus. During this "pregnant" stage, however, the rice grain and plant are most vulnerable to chemicals, so the fungicide would also kill most of the crop.

After he parked his tractor on a dike, Ibrahim called up his neighbor on his iPhone. "Ahmet," he roared, "has a fly gotten in your eye? Can you no longer see your field? You've got blast. It's in the southeast corner, near the imam's house. You really should have monitored your nitrogen levels more carefully."

But Ahmet, Ibrahim told me after hanging up the phone, probably did not need to worry about the rice. "He's a rich man now. He built an apartment in Istanbul—five stories tall. This is on land he bought when he worked in Istanbul, in the past. The rental income from the building is already quite good. Then Kanal Istanbul was unveiled. And the house just happens to be right on the canal path. *Sıfır*: right on the canal."

Kanal Istanbul was an ambitious project to build a fifty-kilometer-long artificial waterway connecting the Black Sea to the Sea of Marmara, bypassing the Bosphorus. The project had been unveiled by AKP prime minister Recep Tayıp Erdoğan in 2011 and had quickly been dubbed his "crazy project." The canal was scheduled to be completed by 2023, in time for the republic's centenary. But as Turkey had entered into a new economic crisis, its completion had been postponed by at least three years.

I had many questions for Ibrahim. How did he know that Ahmet's house was going to be right on the canal?

"Why, he saw the plans," Ibrahim said.

Afterward, I pieced together that Ibrahim had probably been referring to the transportation minister's briefing to the press in January 2018, which had included a map of the proposed route of the canal and the rezoning plans.

I pressed on: How would the value of the apartment increase? Would Ahmet sell it to developers? Or would it be from skyrocketing rental income, instead? What if the house ended up in the path of the canal, instead, and would be seized under eminent domain?

Ibrahim had no answers. What he did know was that *he* had not bought any apartment buildings or land in Istanbul, not even in the nearby town of Bafra. "Instead," he reflected, "I kept buying more land here." He gestured at the sprawling rice fields, now a golden-green and in places punctuated with the yellowish tint of the blast. In many previous conversations, Ibrahim had described his land in the delta as too new and not fertile enough for great rice cultivation. The soil was getting sick from the constant use of agricultural pesticides and fertilizers, which were also flowing in the wetland lakes, where fisheries had already decreased within one generation.

Speculative moral ecologies—even if Ibrahim would not use this term—invite these comparisons of space and time. The rice fields at the edge of the wetland were brought to Istanbul and back in layered assessments of unrealized gains and risks, transmutations of agricultural capital, and increasingly more unpredictable rain patterns: these were worlds suspended in

Figure 9. Rice farmers steeping a pot of tea on a samovar at the edge of the fields, 2018. Photo by Benjamin Siegel.

uncertainty. In the rice fields, the ecology of the blast, pregnant rice, and a new economy of hybrid seeds, pesticide, and land all converged. Speculative conjectures percolated the rice fields together with costly irrigation water, nitrogen, and pesticides.

I return here to the notion of moral ecology as an analytic for environmental precarity in the twenty-first century. The dynamic ecologies of the wetland's shallow water remain at the center of humans', plants', and animals' lived realities of climate change, toxic pollution, and industrial models of agricultural production. Wetlands provide places of hope as we learn to live with rapid sea-level rise, increasingly violent storms and weather, devastating floods, drinking water reservoirs that dry up, islands that disappear underwater, rapid extinction of plants and animal species, and impending catastrophes compounded by pandemics, war, poverty, and inequality. Working through the contested moral ecologies of the wetland helps foreground the heightened stakes wrought of broader changing theaters of life, environment, and livelihood.

The role of wetlands, this book shows, is neither innate nor unchanging. In the early decades of the twentieth century, wetlands came into being as morally charged objects for elite scientists and birders concerned with the survival of birds. Soon the category came to encompass broader preoccupations with the other damaging effects of large-scale drainage and natural resource extraction. Throughout the twentieth century, more than half of the world's marshes, bogs, coastal lagoons, swamps, peat bogs, and other ecologies of shallow water were stabilized, drained, filled, and turned into agricultural fields, cities, and industrial areas. These changes resulted in the loss of habitat for species that had once thrived in seasonal and permanent wetlands, alongside erosion, climatic change, more violent flooding, and groundwater depletion. The wide and all-encompassing category "wetland" emerged as an analytic for scientists and environmentalists to make sense of these shared predicaments of ecological loss.

The Turkish wetland was created in the broader contexts of the republic's changing political economy and vision. Wetland drainage emerged from the legacy of late Ottoman visions of agricultural expansion and population control. Officials of the Turkish republic subsequently undertook the project of turning swamps into wetlands, beginning in the country's capital. Drainage quickly became connected to questions of public health, particularly the eradication of malaria-carrying mosquitoes. The process also helped produce new agricultural lands to resettle exchanged populations, migrants, and refugees. These transformations of land and water carried the imprint of ethnonationalism, simultaneously remaking ecology and Turkishness. Yet as some state officials and scientists pushed the imperatives of drainage, others came to emphasize its destructive effects on Turkish ecology and economy, in the process articulating the Turkish neologism for wetland: *sulakalan*. By the 1980s, Turkish scientists and bureaucrats had recast the nation's wetlands as sites of national value, even as earlier drainage projects continued in earnest.

During the time of my ethnographic fieldwork, from 2012 to 2018, the Turkish wetland had become an even more complicated and multifaceted object. Turkish legislators and officials worked to redefine the scope of wetland management and conservation, at once increasing efforts in the environmental conservation of select wetlands and removing environmental protection in many others. Environmental NGOs celebrated wetlands as sites of biodiversity, cultural heritage, and biocultural hope. University scientists developed new wetland research and education programs, largely grounded in the hydrological and biological sciences and also, increasingly, in social-scientific research. Regional and international collaborations flourished between Turkish and European NGOs, universities, and municipalities on themes of wetland conservation. If the political positioning of individuals and institutions, and disagreements over how wetlands should be managed and for whom, meant that these projects and collaborations were often short-lived, new arrangements would quickly emerge.

Farmers and other rural residents appropriated the term *sulakalan* strategically. Some expressed concern over their loss of access to meadows, lakes, and swamp forests; others lambasted state-led development that once again excluded them and destroyed the environments where they lived and worked; and still others articulated new hope over future economic incentives rooted in sustainable production or tourism in the wetland. Environmental activists connected the ongoing loss of wetlands to a national policy of water expropriation, privatization, and neoliberalism. They connected wetland drainage to the damming of large and small rivers, deforestation, violent population displacement, and the broader erosion of democracy.

The predicaments of the contemporary Turkish wetland remained inextricable from the national politics of water. Beginning in 2008, the national government announced the construction of thousands of run-of-the-river hydropower stations, yoking virtually every stream in the country into larger infrastructures of electricity production—an endeavor enabled by energy-privatization legislation and lucrative contracts that energy firms had drawn up with the government as they rented river access rights and obtained the promise of guaranteed electricity sales. Protesters in many regions faced police and military barriers, even as these mobilizations ultimately proved unable to stop the dams. Yet in the process, water had emerged as an object of political-ecological concern. In their claim making over the putatively proper course of the flow of water from spring to sea, river activists—leftist urbanites and academic intellectuals, alongside union leaders, teachers, retirees, and farmers—connected nature, culture, and politics.

The legibility of the river and its flow as an object of claim making differed from the less scrutable form of the wetland. Ordinary residents, observers, and activists could readily see the rapid destructive effects of small and large dams on local river ecologies and on the social and cultural livelihoods flourishing around rivers. But the disappearance of wetland ecologies proved a more complicated and murky transformation. In a large coastal delta, for example, how does one make sense of the layered effects of underground water extraction, drainage, land filling, industrial and urban pollutants seeping in the water, agricultural chemicals, and new water-thirsty crops? These calculations grow even more intractable when framed against a complicated grid of changing wetland protection and zoning legislation, eminent domain expropriations, and wetland management plans implemented in circumscribed areas. The wetland, never a stable object, seemed even less concrete by comparison.

This book suggests that as different social groups confront and make sense of sweeping infrastructural transformations of ecology and livelihood, they are engaging in broader evaluations of moral ecologies. These are ideas, aspirations, meanings, and evaluations of right and wrong that concern relations between organisms, humans included. Beginning from a moral ecological position requires taking into account the notion that politics is not simply projected *onto* animals, plants, soil, water, sediments, rocks, and other non-human beings and materials. Rather, people make politics *through* them. This book highlights the aspirations, moral relations, and care practices at play in contestations and alliances over environmental change in the wetland.

Wetlands, it is important to add, have not merely reflected or absorbed broader political shifts. I began my research in the weeks preceding the Gezi Park mobilizations of 2013, in a time of increasing grassroots organizing in

Turkey around issues of environmental justice, democratic participation in environmental governance, renewable and sustainable energy, and environmental health. In the winter of 2013, a government corruption scandal resulted in the resignation of several ministers and politicians and increasing strife between the ruling AKP and its former ally, a transnational political, economic, and religious network called the Gülen movement.[1] Over the following months, media outlets fixated on the growing strife between AKP politicians and members of the Gülen movement, involving the dismissal of thousands of prosecutors and police officers and the release of recorded conversations, including a controversial tape recorded between the prime minister and his son as they sought to conceal large amounts of cash in the aftermath of the corruption investigations.

From 2014, a series of new laws tightened control over publications and increased the government's ability to conduct surveillance on any organization seen as constituting a threat to national security. In July 2015, the government commenced new military attacks against the Kurdish Workers' Party (PKK) and instituted curfews that would last for six months in Turkey's Kurdish-majority districts, ending the peace process that had begun in 2013.[2] From July 2016, the government declared a state of emergency lasting two years, during which tens of thousands of public employees were removed from office. In addition, journalists, human rights defenders, and opposition politicians were put on trial and imprisoned under new antiterrorism decrees. New laws restricted the rights of individuals in pretrial detention and restricted lawyers' access to their clients' files. The aftermath of a failed coup attempt in July 2016 led to the increasing repression of Gülen supporters as well as left-wing intellectuals, activists, teachers, and journalists; the suppression of civil liberties and free speech; and the prosecution of academics who had signed peace petitions. These changes coincided as the impact of the Syrian Civil War was felt in Turkey, with the arrival of hundreds of thousands and then millions of refugees.[3]

All this would become directly relevant to my research in unexpected ways. In the context of both war and repression, a number of my collaborators lost their jobs and migrated or sought asylum abroad. A worsening political climate bore directly on the large EU grant for participatory wetland conservation activities on the Kızılırmak Delta, which university professors, environmental educators, and I had written, involving delta farmers in the proposed activities and a collaboration with an Italian NGO. The EU was seemingly unphased by these changes, but as would-be Turkish collaborators found themselves on opposite sides of domestic political debates and some became implicated in the post-coup investigations, the university ultimately turned down the grant.

The spillover of broader politics into the wetland did not decimate its status as an arena for expressing divergent beliefs. Somewhat paradoxically, the wetland in this period became an increasingly neutral space to advance politics by other means. Appeals to notions of moral ecology, I saw, could help advance claims that in other contexts might be politically dangerous or lead to direct repression. Debates over wetland conservation continued, despite increasing authoritarianism. Partly, this resulted from the long-standing understanding of wetlands as objects of primarily technocratic concern, natural areas disconnected from the concerns of formal party politics and rights-based grassroots organizing. It also occurred because of the formal mechanisms of wetland governance: ministerial offices revised five-years plans for wetland conservation, while provincial bureaus published lengthy management plans for each Ramsar wetland and implemented them. Hard copies of the actual management plans did not circulate outside governmental and NGO offices, and wetland residents would never actually see them.

This construction of the wetland as a natural place, or as one of romanticized nature-culture, rendered it semantically open to debates about ecological futures that *appeared* to be removed from human politics. At the same time, the *material* entanglement of wetlands with nearby agricultural, industrial, transportation, and urban infrastructure made it a vital place to advance edgy moral ecological claims. The materiality of wetlands is a recurring theme in this book's discussion of moral and political ecologies, foregrounding the sticky, fluid, muddy, sedimented, elusive materiality of wetlands and showing how the people working and living in wetlands experience, manipulate, and evaluate them.

Materiality, long a central theme in social theory, has enjoyed a recent resurgence in the work of anthropologists and other social scientists. Some have foregrounded the agentive qualities of things. Political theorist Jane Bennett, for example, has argued for the vibrant ontology of matter, ascribing it politics and agency. In Bennett's account, everything is endowed with a swarming materiality, expressing power, and distributed across humans and nonhumans alike.[4] By contrast, anthropologist Stuart McLean foregrounded the capacity of "sticky matter" to assert itself independently of human intention and purpose. Writing against assumptions of objects as bounded and passive, he has taken wetlands, for their interstitial and seeping nature as well as their elusiveness, as an example of "recalcitrant matter": materiality that disrupt projects of modernity.[5] What is swarming and decentered vitality for Bennett is, for McLean, sticky resistance. Both take materiality as a given, an empirical reality.

In response to approaches that have overdetermined matter, including anthropologists who have argued that the essential material properties of water generate cultural meaning,[6] other scholars have instead emphasized the political dimensions of materiality. Political scientist Timothy Mitchell's account of twentieth-century Egyptian politics proposes to take sugarcane, river dams, and malarial mosquitoes as historical actors. In Mitchell's account, binary divisions between natural forces and technoscience emerged as the effects of large-scale infrastructural undertakings. Yet, Mitchell contends, the very possibility of human agency depends on nonhuman elements.[7] Cultural geographer Jessica Barnes has placed the materiality of water front and center in ethnographically tracking how Egyptian farmers, bureaucrats, and scientists seek to control irrigation flows. The technological, hydrological, physical, and agro-ecological processes that determine how, where, and when water moves, Barnes has argued, both enabled and hampered irrigation politics.[8] While Mitchell has taken material properties (e.g., the agentive capacity of river flows or the condition of possibilities of mosquito livelihoods) as a given, Barnes has foregrounded how materials *become* subjects of contrasting agro-ecological politics. Anthropologist of science Stefan Helmreich has suggested that theories about materiality, about the form of water, for example, are also social and political theories and has proposed the study of theory as both an analytic and a phenomenon to be examined.[9]

From my first clumsy steps in deltaic mud during the initial weeks of fieldwork, to attending wetland conferences, talking to wetland bureaucrats, and working with water buffalo farmers and fishers, I could not ignore the distinct and elusive materialities of the lived ecologies we have come to call wetlands. Statements about value, aspirations, and ideals were always grounded in specific material arrangements in the wetland. Fishermen's claims about accessing a fishing lagoon were based on a specific theory of infrastructural work resulting in water flows conducive to thriving fish livelihoods. Educators' attempts to instill a sense of care for the wetland in university students began with stepping into a wetland lake to "feel" the wetland and "become" a water buffalo. Similarly, middle-class residents' engagements with the politics of ecology after the Gezi Park protests of 2013 were also predicated on an embodied, situated, and material sense of their sharing living space with wetland birds and the awareness that the reinforced steel, concrete, and bricks of their houses had also encroached on the nearby marshes. Farmers debated the material effects of roads, canals, water pumps, agricultural chemicals, seeds, and birds in order to make political claims about ecological governance.

All of these were moral ecological claims: my fieldwork interlocutors foregrounded relations between humans and other organisms as constitutive of politics, grounded them in specific material arrangements—including those resulting from infrastructural labor—and connected them to ethical evaluations and aspirations of justice. These moral ecologies can be contrasting, uneven, and violent: the notion does not presume an ethical alignment with my own views, or with the readers'.

The notion of moral ecology foregrounds the critical stakes of how different groups become differentially implicated in, emerge as agents of, and are affected by rapid environmental changes. Moral ecology does not displace but rather keeps political ecology relevant in the twenty-first century as a questioning of received knowledge. In the 1980s and 1990s, political ecologists challenged received assumptions about environmental degradation as the inherent consequences of the livelihood practices of irresponsible smallholders or indigenous people, of rapid population growth, or apolitical ecological process. Instead, they interrogated local power relations centered on resource use, the long-term politics of territorial control, and the effects of national and international markets on localities.[10]

Political ecology continues to offer an essential set of tools in the twenty-first century. Yet its capacities have grown. Over the past fifteen years, scholars have paid increasing attention to the ecological aspects of political ecology by analyzing nonhuman organisms' participation in the political and cultural lives of people, decentering *Anthropos* in history and social theory.[11] To write about the contestations over human livelihood and precarious ecological life in the wetland means also telling stories about water buffaloes, lagoon fish, feral horses, reeds, brine shrimps, little egrets, flamingos, and countless other nonhuman beings that are an inextricable part of our political, social, economic, and cultural lives.

We now face an uncertain future of large-scale geological changes produced by centuries of aggressive extractive capitalism, fossil fuels, and inequality.[12] In Turkey, the materiality of the wetland was not taken for granted as a stable starting point, despite countless representations in environmental NGO reports of the wetland as a natural infrastructure, a sponge, or the kidneys of the earth. Rather, it emerged as the subject of debates, contestations, imaginations, and struggles. Materiality is implicated in local theories of ecological possibility: wetlands are drying and dying mudlands in the interstices of other infrastructures, flood surge barriers for sea-level rise, water reservoirs for increasing drought, ecologies sustained through irrigation infrastructure, infrastructural form for the ecomodern cities of the future, or material inspiration for reclaiming nature-culture traditions. Each way to make a wetland is also a normative and aspirational claim for justice and intersubjectivity constituted through relations with other organisms, materials, and temporality: a moral ecology.

Notes

Introduction

1. Throughout this book, I use pseudonyms and have changed several toponyms and institutional names to protect my interlocutors' identities. All translations of Turkish dialogues and documents are my own.

2. UNESCO, "The Criteria for Selection."

3. Derneği, "Flamingo Bölgesine Otoban."

4. "İzmir'in Gediz Deltası UNESCO Dünya Doğa Mirası Ilan Edilsin!"

5. Ruzicka, *Trouble in the West*, 97–98.

6. Choy, *Ecologies of Comparison.* ← problematic footnote

7. Anthropologists have written on amphibious spaces, in contrast to visions that privilege terraforming and divisions between land and water, as those where social life and infrastructure are organized around water flows. Krause, "Rhythms of Wet and Dry"; Richardson, "Terrestrialization of Amphibious Life."

8. T. Hughes, "Evolution of Large Technological Systems"; Star, "The Ethnography of Infrastructure."

9. Adalet, *Hotels and Highways.*

10. Carse, "Infrastructure."

11. Björkman, *Pipe Politics, Contested Waters.*

12. Larkin, *Signal and Noise.*

13. Schwenkel, "Spectacular Infrastructure."

14. Farmer, "Willing to Pay."

15. Anand, *Hydraulic City*; Von Schnitzler, "Traveling Technologies."

16. Nucho, *Everyday Sectarianism in Urban Lebanon.*

17. Meiton, *Electrical Palestine.*

18. Elyachar, "Phatic Labor"; Simone, "People as Infrastructure."

19. Kurtiç, "Sedimented Encounters"; Stamatopoulou-Robbins, *Waste Siege*.

20. Blackbourn, *The Conquest of Nature*; Yeh, "Nature and Nation in China's Tibet"; Pritchard, *Confluence*; Ritvo, *The Dawn of Green*; R. White, *The Organic Machine*.

21. Costanza, Farber, and Maxwell, "Valuation and Management of Wetland Ecosystems."

22. Carse, "Nature as Infrastructure"; Morita, "Multispecies Infrastructure."

23. Bruun Jensen, "Amphibious"; Cons, "Staging Climate Security"; Hetherington, "Keywords of the Anthropocene"; Kim, "Toward an Anthropology of Landmines"; Stoetzer, "Ruderal Ecologies"; Tsing, *Mushroom at the End of the World*.

24. Millar, *Reclaiming the Discarded*, 100.

25. In the last two decades, anthropologists have elaborated on notions and practices of economic morality in great detail, inspired by the earlier work of E. P. Thompson and James Scott, which emphasized communal understandings of redistributive economic justice in the face of profiteering and exploitation and taking part in a growing interest on the phenomenology, ontology, and politics of moral and ethical worlds. Mattingly and Throop, "Anthropology of Ethics and Morality"; Scott, *Moral Economy of the Peasant*; Thompson, "The Moral Economy." I propose that the moral stakes of ecological relations also be critically examined.

26. Carse, "Nature as Infrastructure"; Stoetzer, "Ruderal Ecologies"; Anand, Gupta, and Appel, *The Promise of Infrastructure*.

27. Scott, *Moral Economy of the Peasant*; Muehlebach, *The Moral Neoliberal*; Thompson, "The Moral Economy."

28. Baker et al., "Mainstreaming Morality"; Dove and Kammen, "Epistemology of Sustainable Resource Use"; Martinez-Reyes, *Moral Ecology of a Forest*.

29. Rizvi, "Moral Ecology of Colonial Infrastructure"; Campbell, "Moral Ecologies of Subsistence."

30. Erensü and Alemdaroğlu, "Dialectics of Reform and Repression."

31. Adalet, *Hotels and Highways*.

32. Dolbee, "Desert at the End of Empire"; Mikhail, *Nature and Empire*; Özkan, "Remembering Zingal"; S. White, "Rethinking Disease in Ottoman History"; S. White, *Climate of Rebellion*.

33. Erensü, "Powering Neoliberalization"; Erensü and Karaman, "Work of a Few Trees."

34. Arsel, Akbulut, and Adaman, "Environmentalism of the Malcontent."

35. Knudsen, "Protests against Energy Projects."

36. Evren, "Rise and Decline of an Anti-dam Campaign."

37. Erensü, "Fragile Energy."

38. "Kızılırmak Deltası, Gediz Deltası"; Arı, "Visions of a Wetland," 25–26.

39. IUCN, IUBP, and IWRB, *Proceedings*, 51–52.

40. IUCN, IUBP, and IWRB, 39–40.

41. Magnin and Yarar, *Important Bird Areas in Turkey*, 8; Ertan, Kılıç, and Kasparek, *Türkiy'nin Önemli Kuş Alanları*.

42. Magnin and Yarar, *Important Bird Areas in Turkey*, 293–296.

43. Moser, Prentice, and Van Vessem, *Waterfowl and Wetland Conservation in the 1990s*, 166.

44. *Water for People, Water for Life*.

45. "Marmara Denizi'nin 2 Katı."

46. Geylan, "Sulak Alanların Kuruduğu Algısı."

47. Akçam, *Young Turks' Crime against Humanity*; Goffman, *Izmir and the Levantine World*; Hovannisian, *Armenian*; Meichanetsidis, "Genocide of the Greeks"; Milton, *Paradise Lost: Smyrna*; Popov, *Culture, Ethnicity and Migration*.

48. Here I understand networks broadly, following Bruno Latour, as the work of tracing connections between things and the traces left behind by actors, but I also agree with Annalise Riles's "anti-network" analysis, which emphasizes opacity, ontological multiplicity, and disconnections and the work of creating and manipulating legitimacy. Latour, *Reassembling the Social*; Riles, "The Anti-network."

49. Choy, *Ecologies of Comparison*.

50. Choy.

51. Ash, *Draining of the Fens*; Blackbourn, *The Conquest of Nature*; Biggs, *Quagmire*; Mathur and da Cunha, *Mississippi Floods*; Guarasci, "The National Park"; Husain, "In the Bellies of the Marshes"; Snowden, *The Conquest of Malaria*.

52. Ahram, "Development, Counterinsurgency"; Gratien, "The Ottoman Quagmire."

53. Bozdogan, *Modernism and Nation Building*; Evered, "Draining an Anatolian Desert."

54. Wilson, *Seeking Refuge*.

55. Star and Griesemer, "Institutional Ecology."

Chapter 1

1. An office located within the Ministry of Energy and Natural Resources.

2. While the Turkish government frequently distributed reclaimed land to landless farmers and immigrants, the country never undertook a program of land reform. See Karaömerlioğlu, "Elite Perceptions of Land Reform"; Parvin and Hic, "Land Reform versus Agricultural Reform."

3. IUCN, IUBP, and IWRB, *Proceedings*, 13–14, 52.

4. In contemporary Turkey, agricultural engineers, water-works officials, and urban residents continue to use the term *bataklık*. They choose to use the term *sulakalanlar* with a positive connotation to describe biodiverse ecologies, sites of thriving human and nonhuman livelihoods and biocultures, field sites for scientific research, and environments worthy of protection. By contrast, *bataklar* denote problematic environments: unruly or hazardous nature that threatens or resists national progress, settlement, and human livelihood. Farmers, fishers, and other rural residents living and working in wetland areas often use neither term. Rather, they have more

place-specific names describing different kinds of watery places, in different seasons, and denoting different uses.

5. I mean both the invention of particular ways of looking at, measuring, and classifying wetlands, inspired by Dilip da Cunha's work on rivers, and the rise of an environmentalist imperative to protect them, as in Harriet Ritvo's work on environmentalist arguments for the preservation of watery landscapes. Da Cunha, *The Invention of Rivers*; Ritvo, *The Dawn of Green*.

6. Haraway, "Teddy Bear Patriarchy"; D. Hughes, "Water as Boundary"; Igoe and Brockington, "Neoliberal Conservation"; Jacoby, *Crimes against Nature*; Neumann, *Imposing Wilderness*; Peluso, *Rich Forests, Poor People*; Walley, *Rough Waters*; West, *Conservation Is Our Government Now*; West, Igoe, and Brockington, "Parks and Peoples."

7. Andersson, "Environmentalists with Guns."

8. Clark, *Twice a Stranger*; Gökay and Aybak, "Identity, Race and Nationalism in Turkey"; Kadioğlu, "Paradox of Turkish Nationalism"; Üngör, *Making of Modern Turkey*; Zurcher, *Turkey*; J. White, *Muslim Nationalism and the New Turks*.

9. Gratien, "The Ottoman Quagmire," 589.

10. Gratien, 590.

11. Gratien, 590.

12. Evered and Evered, "Governing Population," 475–476; Evered and Evered, "A Conquest of Rice."

13. Ahram, "Development, Counterinsurgency."

14. Koylu and Doğan, "Sıtma Mücadelesi."

15. Evered and Evered, "Governing Population."

16. Evered, "Draining an Anatolian Desert."

17. Yılmaz, "Altınkaya Barajı'nın Vezirköprü'ye Etikleri."

18. "Bataklıkların Kurutulması ve Bundan Elde Edilecek Topraklar Hakkinda Kanun."

19. Asif and Taneja, "Animals, Ethics, and Enchantment"; Erman, "Bird Houses"; Faroqhi, *Animals and People*.

20. Belon, *L'histoire de la nature des oyseaux*.

21. McGhie, *Henry Dresser*, 108.

22. Dresser, "On a Collection of Birds from Erzeroom."

23. Kirwan et al., *The Birds of Turkey*.

24. Turner, "Thomas Robson," 46.

25. Kirwan et al., *The Birds of Turkey*.

26. Wahby, "Les oiseaux de la region de Stamboul," 174.

27. The terms "wildfowl" and "waterfowl" mostly refer to swans, ducks, and geese, much prized by hunters. The term "waterbird," which has been used since the 1950s and eventually replaced the categories "wildfowl" and "waterfowl," refers to birds that live on or around bodies of water.

28. Powell, *Vanishing America*.

29. Errington, *Of Men and Marshes*.

30. Doughty, *Feather Fashions*.

31. Moore-Colyer, "Feathered Women and Persecuted Birds"

32. Bruyninckx, *Listening in the Field*; Dunlap, *In the Field*; Jacobs, *Birders of Africa*.

33. Named after ornithologist and painter John James Audubon.

34. Wilson, *Seeking Refuge*.

35. Palmer, *Legislation*, 20.

36. Wilson, *Seeking Refuge*.

37. In another volume, Berry summarized the factors responsible for the decline of home-bred wild ducks in Scotland: natural enemies (crows, gulls, swans, rats, swans, angler fish, pike), oil pollution, stake netting, flight netting, egg taking, excessive shooting, road traffic, boats and shipping, airplanes, agriculture, natural feeding, drainage and afforestation, and the effects of climate. Berry, *International Wildfowl Inquiry*, 2:180.

38. Berry, 1:7.

39. Berry, 1:4.

40. Berry, 1:8. On the transformation of water into a subject of hydrological knowledge, see Carroll, "Water and Technoscientific State Formation in California"; Linton, *What Is Water?* On soil erosion and the "dust bowl" as an outcome of white settler colonialism and the racialized transformation of economic networks, see Holleman, *Dust Bowls of Empire*.

41. IUCN, IUBP, and IWRB, *Proceedings*, 85; Reisman, *Turkey's Modernization*.

42. Anthropological scholarship on international institutions has largely focused on development apparatus: for example, Ilcan and Phillips, "Developmentalities and Calculative Practices"; Keck and Sikkink, *Activists beyond Borders*; Staples, *The Birth of Development*.

43. Carp, *Directory of Wetlands*, 1.

44. *First Annual Report of Severn Wildfowl Trust*.

45. Kumerloeve, *Bibliographie*.

46. Kumerloeve.

47. Nowak, *Biologists in the Age of Totalitarianism*.

48. Niethammer, "Die Vogelwelt von Auschwitz."

49. For example, Beaman, Porter, and Vittery, *Bird Report 1970–1973*, 289–301; Kirwan et al., *The Birds of Turkey*, 115.

50. Watson, "Aegean Bird Notes 1."

51. Kitson and Porter, *Bulletin No. 5*, 3.

52. Hagen, *An Entangled Bank*.

53. Cézilly and Salathé, "Luc Hoffmann."

54. "[Wetlands] are generally composed of habitats which are truly aquatic, those which are only moist, and those which are truly terrestrial. Interesting edge effects are displayed on the limits, including the possibilities of amphibious life. Moreover, seasonal cycles often result in the same place being aquatic for one part of the year and terrestrial for the other. Aquatic habitats are in a delicate balance and undergo very

quick evolution. Changes of water level or quality at once have profound effects on the communities and, as they can often be controlled without great difficulty permit experimental research in natural areas." Hoffmann, "Research," 3:400.

55. Hoffmann, 3:402.

56. On ecosystems as energy transactions, see Hagen, *An Entangled Bank*; Martin, "Proving Grounds."

57. Hoffmann, "Research," 3:402.

58. On the mutual constitution of scientific laboratories and field sites, see Kohler, *Landscapes and Labscapes*.

59. Hoffmann, "Research," 3:406.

60. Matthews, *Ramsar Convention on Wetlands*.

61. Cézilly and Salathé, "Luc Hoffmann."

62. *Project Mar*, 1:24–25.

63. *Project Mar*, 1:61. On wetland conservation and classification in the United States, see Shaw and Fredine, *Wetlands of the United States*.

64. *Project Mar*, 1:201–202.

65. *Project Mar*, 1:38–39.

66. *Project Mar*, 1:46.

67. Star and Griesemer, "Institutional Ecology."

68. *Project Mar*, 1:26–30.

69. *Project Mar*, 1:10.

70. See "George Atkinson-Willes"; and "In Memoriam: George Atkinson-Willes."

71. Atkinson-Willes, *Liquid Assets*, 1.

72. Atkinson-Willes, 1.

73. Atkinson-Willes, 1.

74. Olney, *Project Mar*, 2:86.

75. Olney, 2:11. Communication studies scholar Edward Schiappa suggested that definitions of the wetland are at once ethical and political: reflecting normative understandings and aspirations and becoming stabilized through persuasion or coercion. Schiappa, "Towards a Pragmatic Approach to Definition."

76. Anthropologists and historians of science have argued that definitions like those of the wetlands are not stable but rather travel as political objects at different scales. Further, these definitions are shaped by the moral visions of the people who craft them, appropriated for strategic and contingent goals, and transformed by material environments and layered histories of work and infrastructure. See Hecht, *Being Nuclear*; Helmreich, *Alien Ocean*; Lewis, *Inventing Global Ecology*; Mavhunga, *Transient Workspaces*.

77. Olney, *Project Mar*, 2:18.

78. Olney, 2:8.

79. Olney, 2:9.

80. Meyer, "When Dismal Swamps Became Priceless Wetlands."

81. In the 1990s, these would be reframed as "ecosystem services." See Costanza et al., "Value of the World's Ecosystem Services."

82. In the United Kingdom, the Netherlands, and the Soviet Union. See Kuijken, "A Short History of Waterbird Conservation."

83. IUCN, IUBP, and IWRB, *Proceedings*, 13.

84. IUCN, IUBP, and IWRB, 259.

85. His job was at the General Directorate of Mineral Research and Exploration (MTA). "Tansu Gürpinar'la Söyleşi."

86. IUCN, IUBP, and IWRB, *Proceedings*, 88.

87. Carp, *Final Act and Summary Record*; De Klemm and Créteaux, *Legal Development of the Ramsar Convention*; Matthews, *The Ramsar Convention*.

88. In 1995, the IWRB merged with two organizations established in the 1980s, the Asian Wetland Bureau and Wetlands for the Americas, and became Wetlands International. Today, Wetlands International is headquartered in the Netherlands and active on projects of worldwide wetland conservation that intersect with development and human health concerns. The stabilization of the wetland was enabled by this shape-shifting institutional continuity.

89. The Ramsar Bureau, an independent organization, was established in 1993 and headquartered in Switzerland in the same office as the IUCN.

90. EPFT, *The Wetlands of Turkey*, 5, 171.

91. Özesmi, Somuncu, and Tunçel, "Sultan Sazlığı Ekosistemi."

92. Özesmi, Somuncu, and Tunçel, 275–288; Brooks, *Sandgrouse*.

93. Erdem, *Sulak Alanların Önemi*; Erdem, *Turkey's Bird Paradises*.

94. Navid, "Letter to M. Paszkowski."

95. Erdem, *Turkey's Bird Paradises*, 3.

96. Erdem, 34–43.

97. Davies and Claridge, *Wetland Benefits*.

98. Law No. 25818, subsequently revised in May 2005 and April 2014.

99. Doğal Hayatı Koruma Derneği, *Göksu Delta*; Doğal Hayatı Koruma Derneği, *Towards Integrated Management in the Göksu Delta*.

100. Gürpinar, "Doğal Çevre Ve Fotoğraf."

101. Gürpinar.

102. Davis, *Environmental Imaginaries*.

103. Adaman, Hakyemez, and Özkaynak, "Political Ecology of a Ramsar Conservation Failure"; Adaman and Arsel, *Environmentalism in Turkey*; Arsel, Akbulut, and Adaman, "Environmentalism of the Malcontent"; Güçlü and Karahan, "A Review"; Evered, "Draining an Anatolian Desert"; Karadeniz, Tırıl, and Baylan, "Wetland Management in Turkey"; Ignatow, "Economic Dependence and Environmental Attitudes"; Şekercioğlu et al., "Turkey's Globally Important Biodiversity in Crises."

Chapter 2

1. Zandi-Sayek, *Ottoman Izmir*.

2. Strang, *The Meaning of Water*; Orlove and Caton, "Water Sustainability."

3. Culutral geographer Jessica Barnes has called ethnographic attention to how farmers, engineers, and bureaucrats remake agricultural water materialities at different scales through technology, bureaucracy, scientific claims, and labor. Barnes, *Cultivating the Nile*.

4. Anthropologists Tanya Richardson and Gisa Weszalnys analyze historical "assumptions about what resource substances are, their affordances and what sustains them." Richardson and Weszkalnys, "Introduction: Resource Materialities," 18–19.

5. Richardson, "Where the Water Sheds"; Kurtiç, "Sediment in Reservoirs."

6. Choy, *Ecologies of Comparison*; Doane, *Stealing Shining Rivers*; Lowe, *Wild Profusion*; West, *Conservation Is Our Government Now*; Walley, *Rough Waters*.

7. For example, Baker et al., "Mainstreaming Morality"; Dove and Kammen, "Epistemology of Sustainable Resource Use"; Martinez-Reyes, *Moral Ecology of a Forest*.

8. Gewertz and Errington, "Doing Good," 19–20; Gewertz and Errington, "Pheasant Capitalism."

9. Raffles, *In Amazonia*; Barnes, *Cultivating the Nile*; Helmreich, "Nature/Culture/Seawater"; Kurtiç, "Sedimented Encounters"; Ballestero, *A Future History of Water*; Ballestero, "The Anthropology of Water."

10. Orlove and Caton, "Water Sustainability"; Strang, *The Meaning of Water*.

11. Ballestero, "Ethics of a Formula"; Linton, *What Is Water?*

12. Helmreich, "Nature/Culture/Seawater."

13. Giblett, *Postmodern Wetlands*; McLean, "Black Goo"; Morita, "Infrastructuring Amphibious Space"; Richardson, "Terrestrialization of Amphibious Life."

14. Krause, "Reclaiming Flow," 90.

15. Richardson, "Terrestrialization of Amphibious Life."

16. "Dünyanın 3. Büyük Flamingo Adası."

17. Ernoul and Wardell-Johnson, "Governance in Integrated Coastal Zone Management"; Finlayson et al., "The Ramsar Convention."

18. Ernoul, "Entre Camargue et Delta de Gediz."

19. Erol, *Ottoman Crisis in Western Anatolia*.

20. Egemen, *Türkiyede Tuzculuk ve Çamaltı Tuzlası*.

21. Erol, *Ottoman Crisis in Western Anatolia*.

22. Saltan and Okar, *Summary of the Salt Industry*.

23. Sıkı, "Izmir Kuşcenneti'nin Dünü, Bügünü, ve Yarını."

24. Sıkı and Baran, "Çamaltı Tuzlası'ndaki 'Kuş Cenneti,'" 4–5.

25. Özkan, "Remembering Zingal."

26. Sıkı, "Izmir Kuş Cennetinin Tarihçesi."

27. Sıkı; Sıkı, "Izmir Kuşcenneti'nin Dünü, Bügünü, ve Yarını."

28. Latour and Woolgar, *Laboratory Life*.

29. According to anthropologist Cristina Grasseni, "skilled vision," among northern Italian breeders and cattle farmers, is a learned way of seeing that is culturally, socially, and materially inflected. It is a mode of perception that also leads to the inscription of expert judgment (of cow breeds, in Grasseni's ethnography, and of habitat boundary, in my example here) on paper. See Grasseni, "Skilled Vision." Heather

Paxson advances a similar argument as she notices that in the United States artisanal cheesemakers practice a "tacit ability to gather and interpret sensory data" in an "ecology of activity that scales from the microbial to the human." This tacit ability is sensorial, based on personal experience, even as it scales up to production that fits regulatory standards and the bureaucratic structures of national food markets. See Paxson, "Post-Pasteurian Cultures."

30. Kohler, "Paul Errington, Aldo Leopold, and Wildlife Ecology."

31. Choy, *Ecologies of Comparison*, 26.

32. Kaplan and Hepcan, "Examination of Ecological Risk Assessment."

33. Sulak Alanlar Şubesi Müdürlüğü, *Gediz Deltası Sulakalan Yönetim Planı*.

34. Russi et al., *Economics of Ecosystems and Biodiversity*.

35. Costanza, Farber, and Maxwell, "Valuation and Management of Wetland Ecosystems"; Costanza et al., "Value of the World's Ecosystem Services"; Groot, Stuip, and Finlayson, *Guidance for Valuing the Benefits*; Finlayson et al., "The Ramsar Convention"; Jax et al., "Ecosystem Services and Ethics."

36. On the history and practices of participatory conservation, see Walley, *Rough Waters*; West, *Conservation Is Our Government Now*; Igoe and Brockington, "Neoliberal Conservation."

37. Sönmez and Onmuş, *Gediz Deltası Yönetim Planı*.

38. Adalet, *Hotels and Highways*.

39. Darkot, "Ege Haliçlerinin Mense ve Tekamülü," 37.

40. Öktem, "The Nation's Imprint," 19; Sahakyan, *Turkification of the Toponyms*.

41. Darkot, "Ege Haliçlerinin Mense ve Tekamülü," 94.

42. Darkot, 39.

43. "Kuşlar Kurtuldu Sıra Balıklarda."

44. "İzmir Körfezi'ni Bekleyen Büyük Tehlike."

45. Bhattacharyya, *Empire and Ecology in the Bengal Delta*.

46. Larkin, *Signal and Noise*; Larkin, "Politics and Poetics of Infrastructure"; Simone, "People as Infrastructure"; Carse, "Nature as Infrastructure"; Von Schnitzler, "Traveling Technologies"; Harvey and Knox, *Roads*.

47. Pelkmans, "Social Life of Empty Buildings."

48. Sulak Alanlar Şubesi Müdürlüğü, *Gediz Deltası Sulakalan Yönetim Planı*.

49. On feral horse sociality, see Rees, *Horses in Company*.

50. *Technical Assistance for the Integrated Management of Wetlands*.

51. *Technical Assistance for the Integrated Management of Wetlands*, 19.

52. Doğa Derneği, "Gediz'in Vazgeçilmesi."

53. For arguments against dichotomies of wild/domestic, see O'Connor, "Working at Relationships"; Cassidy and Mullin, *Where the Wild Things Are Now*.

54. Gündoğdu, "The State and the Stray Dogs"; Mikhail, "A Dog-Eat-Dog Empire"; Yıldırım, "Hayvan Tecritinin Dışı ve Ötesi."

55. Bayram, "İzmir Kuş Cenneti'nde."

56. Lowe, "Making the Monkey."

57. Takacs, *The Idea of Biodiversity*.

58. Errington and Gewertz, "Managing an Endangered Species."
59. For example, Hathaway, *Environmental Winds.*
60. Van Dooren, *Flight Ways.*

Chapter 3

1. An earlier version of this chapter appears in Scaramelli, "The Delta Is Dead."
2. Barnes and Alatout, "Water Worlds" ; Richardson, "Where the Water Sheds"; Yates, Harris, and Wilson, "Multiple Ontologies of Water."
3. Biggs, *Quagmire*; Gruppuso, "Edenic Views in Wetland Conservation"; Guarasci, "The National Park"; Husain, "Changes in the Euphrates River"; Özesmi, "Ecological Economics of Harvesting Sharp-Pointed Rush."
4. Hetherington, "Keywords of the Anthropocene."
5. Bubandt and Tsing, "Feral Dynamics of Post-industrial Ruin."
6. Anthropologist Kalyanakrishnan Sivaramakrishnan has called attention to histories of religious identity, resource management, and politics that subtend what he calls "ethics of nature:" attachments nourished through practice that combine utilitarian concerns and emotional ties. Sivaramakrishnan, "Ethics of Nature," 1263.
7. Adaman, Akbulut, and Arsel, *Neoliberal Turkey and Its Discontents*; Erensü and Karaman, "The Work of a Few Trees."
8. Fırat, "The Most Eastern of the West"; Erensü, "Powering Neoliberalization."
9. Knudsen, "Protests against Energy Projects in Turkey."
10. Arsel, Akbulut, and Adaman, "Environmentalism of the Malcontent"; Knudsen, "Protests against Energy Projects in Turkey"; Voulvouli, *From Environmentalism to Transenvironmentalism.*
11. Ayfer Bartu Candan and Kiray Kolluoğlu argued that the new urban forms of the middle-class gated communities and housing for the poor are interconnected. By contrast, Kimberly Hart argued that the residents of an Aegean village that became a suburban neighborhood of wage workers still understood themselves as independent from the state provision of infrastructure. Bartu Candan and Kolluoğlu, "Emerging Spaces of Neoliberalism"; Hart, "Suburbanization of Rural Life."
12. Scaramelli, "Fish, Flows, and Desire in the Delta."
13. Datta, "Gendered Nature and Urban Culture."
14. Coleman, *A Moral Technology.*
15. Simone, "People as Infrastructure."
16. Igoe and Brockington, "Neoliberal Conservation."
17. Rangan, *Of Myths and Movements.*
18. Errington and Gewertz, "Managing an Endangered Species"; Deborah Bird-Rose, "Flying Fox"; Kim, "Toward an Anthropology of Landmines"; Cons, "Staging Climate Security."
19. Raymond Williams understood emergent social and cultural practices as opposed to dominant orders, in a dialectic of incorporation and resistance. Michael Fischer theorized emergent forms of life in late capitalism as social mediations of science. Emergent,

in the sense of processes arising unexpectedly and distinct from existing conditions, often carries a normative connotation: for example, Eban Kirksey has interpreted emergent ecologies as symbiotic associations of living creatures that disrupt the existing order while generating new possibilities for mutual flourishing. See Williams, "Dominant, Residual, and Emergent"; Fischer, "Emergent Forms of Life"; Kirksey, *Emergent Ecologies*. Regarding emergent ecologies that thrive in infrastructural rubble and ruins, see Stoetzer, "Ruderal Ecologies"; Tsing, *Mushroom at the End of the World*.

20. Jensen, "The *Umwelten* of Infrastructure."

21. Anthropologists have theorized environmental infrastructure as the managerial notion that ecologies themselves perform the work of human-built systems, foregrounding the entanglement of the built and natural environment. See Carse, *Beyond the Big Ditch*; Hetherington, *Infrastructure, Environment*. For instance, as floating rice supports the Chao Pharaya Delta's irrigation system, engineers and environmental managers envision multispecies and natural infrastructures. See Morita, "Infrastructuring Amphibious Space."

22. Anthropologist Joe Masco has argued that species once understood as vulnerable to nuclear contamination have become icons of ecological survival and purity in remediation zones.

23. Blackbourn, *The Conquest of Nature*; Pritchard, *Confluence*; Ritvo, *The Animal Estate*; R. White, *The Organic Machine*.

24. Moral ecologies also emerge from embodied practices of cultivation of place, identity, and belonging, sometimes rooted in the land, or flowing (and sedimented) in moving waters.

25. Edelman, "Bringing the Moral Economy Back."

26. Thompson, "The Moral Economy."

27. Scott, *Moral Economy of the Peasant*. Scott's interpretation added attention to specific values, emotions, and senses of justice to Thompson's focus on traditional norms.

28. Hart, *And Then We Work for God*.

29. Thompson, "The Moral Economy," 79.

30. Polanyi, *The Great Transformation*.

31. Muehlebach, *The Moral Neoliberal*, 20. Similarly, Heather Paxson has argued that artisanal cheesemakers in the United States staked moral claims about their work that were grounded in the specifics of their production ecologies and within economic and agricultural systems. Their moral practices involved working with multispecies collaborators, (post)industrial technology, tacit knowledge, and market networks. Paxson, *The Life of Cheese*.

32. Baker et al., "Mainstreaming Morality"; Martinez-Reyes, *Moral Ecology of a Forest*.

33. Dove and Kammen, "Epistemology of Sustainable Resource Use"; Martinez-Reyes, *Moral Ecology of a Forest*.

34. Agrawal, *Environmentality*. Moral ecologies are also embedded in broader political ecologies.

35. Erol, *Ottoman Crisis in Western Anatolia*.

36. Goffman, *Ottoman Empire and Early Modern Europe*.

37. Goffman; Zandi-Sayek, *Ottoman Izmir*.

38. Milton, *Paradise Lost*; Meichanetsidis, "Genocide of the Greeks."

39. Neyzi, "Remembering Smyrna/Izmir."

40. Yeh, "Nature and Nation in China's Tibet."

41. Doğer, *Ilk Insanlardan Yunan Işgaline Kadar Menemen*.

42. Alpbaz, "Homa Dalyanı."

43. Harvey and Knox, *Roads*.

44. In the 1990s, scientific research in Homa Lagoon focused on the role of phytoplankton on fish populations. In the 2000s, phytoplankton itself became the subject of research. See Yazıcı and Büyükışık, "Homa Dalyanı."

45. Kızılkaya, "Homa Dalyanı'nın Yeni Yüzü"; Karademir Erol, "Körfezin Son Dalyanı Homa Koruma Altında."

46. Paxson and Helmreich, "Perils and Promises of Microbial Abundance."

47. Gültekin, "Flamingolar Yaşasın Diye Her Cuma Eylem."

48. Zengin, "The Afterlife of Gender."

49. "Suriyeli Çocuklardan Temsili Cenaze Töreni"; "Manisalı Üretici Üzümün Cenaze Namazını Kıldı"; "Kararname Ekleyen Kamu Müteahhitleri, Temsili Cenaze Namazı Kıldı."

50. Özkırımlı, *Making of a Protest Movement in Turkey*; Tambar, "Brotherhood in Dispossession."

51. Erensü and Karaman, "Work of a Few Trees."

52. Parla, "Critique without a Politics of Home?"

53. The wetland legislation of 2005 was amended in 2010 and 2014. See: "Sulak Alanların Korunması Yönetmenliği."

54. From the Office of Wetlands (Sulak Alanlar Şube Müdürlüğü), a subdivision of the Nature Protection and National Parks Directorate, within the Ministry of Water and Forestry.

55. Yıldırım went on to become the minister of transport, maritime affairs, and communications; AKP party leader; and then prime minister, until the office was abolished in the referendum of June 2018.

56. "Hedef 2023."

57. Schwenkel, "Spectacular Infrastructure."

58. Farmer, "Cairo Ecologies."

59. Anand, "Pressure."

60. Von Schnitzler, "Traveling Technologies."

61. Arif Ali Cangı and Cem Altparmak on behalf of Egeçep et al., Request for Injunction, filed at Izmir District Court, May 4, 2017.

62. Five years earlier, flamingos had been at the center of an environmental NGO's mobilization against the proposed use of the wetland areas as a site to deposit mud dredged from the bottom of Izmir Bay. See Özkırlı, "Flamingolar Çamura Gömülmesin."

63. In contrast with Takacs, *The Idea of Biodiversity.*
64. Karakoyun, "İzmir Körfez Geçişi'ne Durdurma."
65. Egeçep et al. v. Ministry of Environment and Urbanization (Izmir 3rd District Court, November 30, 2018); Aktay, "İzmir Körfez Geçişi Projesi İptal Kararı İzmir Halkına Armağandır."
66. See also Ramsar Convention Secretariat, *Addressing Change,* 19–24.
67. Howe et al., "Paradoxical Infrastructures," 559.
68. Harvey and Knox, "The Enchantments of Infrastructure."
69. Thompson, "The Moral Economy."

Chapter 4

1. An earlier version of this chapter appears in Scaramelli, "The Wetland Is Disappearing."
2. I lived in the Kızılırmak Delta in 2013 and 2015 and then again in the summers of 2017 and 2018.
3. On the making of natural resources, see Richardson and Weszkalnys, "Introduction: Resource Materialities"; Robbins, *Political Ecology.*
4. Scientific sense making of the world is embodied practice and ethical positioning and mutually constituted with notions of livelihood and multispecies and ecology. See Scaramelli, "Making Sense of Water Quality"; Scaramelli, "The Delta Is Dead."
5. Alkan, "Bilim ve Teknoloji Çalışmaları'nda Altyapılar," 105–124; Dalyan, "Latent Lives"; Mutlu, "Transnational Biopolitics and Family-Making in Secrecy"; Kayaalp, *Remaking Politics, Markets, and Citizens in Turkey*; Kasdoğan, "Potentiating Algae, Modernizing Bioeconomies"; Kurtiç, "Sediment in Reservoirs"; Zeybek, "Türkiye'de Endüstriel Hayvancılığın Seyri."
6. Dole, *Healing Secular Life*; Önder, *We Have No Microbes Here*; Sanal, *New Organs within Us.*
7. Alber and Drotbohm, *Anthropological Perspectives on Care*; Buch, "Anthropology of Aging and Care"; Mol, *The Logic of Care*; Barnes and Taher, "Care and Conveyance."
8. Gagné, *Caring for Glaciers,* 7.
9. Tronto, *Moral Boundaries,* 103.
10. Bellacasa, "Making Time for Soil"; Kirksey, *Emergent Ecologies*; Martin, Myers, and Viseu, "The Politics of Care in Technoscience."
11. Bellacasa, "Making Time for Soil"; Latour, "Why Has Critique Run out of Steam?"
12. Caring for ecologies, like caring for objects or relationships, entails practices of knowing, organizing, classifying, and selecting; it is telling that to "care" and to "curate" are etymologically connected in English and in Romance languages.
13. Cronon, "The Trouble with Wilderness"; Lowe, *Wild Profusion*; Moore, *Suffering for Territory*; Walley, *Rough Waters*; West, *Conservation Is Our Government Now.*

14. Abu El-Haj, *Facts on the Ground*; Bartu-Candan, "Remembering a 9000 Years Old Site"; Lorimer, *Wildlife in the Anthropocene.*

15. Erensü and Karaman, "Work of a Few Trees."

16. Harris, "Contested Sustainabilities."

17. Harris, "Irrigation, Gender."

18. Aksu, Erensü, and Evren, *Sudan Sebepler*; Erensü, "Abundance and Scarcity"; Kurtiç and Kadirbeyoğlu, "Problems and Prospects."

19. Evren, "Rise and Decline of an Anti-dam Campaign."

20. Arsel, Akbulut, and Adaman, "Environmentalism of the Malcontent"; Knudsen, "Protests against Energy Projects in Turkey."

21. Strabo of Amaseia, *Complete Works of Strabo*, 695.

22. Yılmaz, *Bafra Ovası'nın Beşeri ve İktisadi Coğrafyası.*

23. Özesmi, "Conservation Strategies"; Quataert, *Social Disintegration and Popular Resistance*; Quataert, *Consumption Studies.*

24. Kayaalp, *Remaking Politics.*

25. Köseoğlu, "Nüfus Kaynakları ve Sözlü Tarih."

26. Akçam, *Young Turks' Crime against Humanity*; McCarthy, *Muslims and Minorities*; Ekmekcioglu, *Recovering Armenia*; Zurcher, *Turkey*; Prott, *Politics of Self-Determination.*

27. Clark, *Twice a Stranger*; Kirişci, "Migration and Turkey," 266–300; Meichanetsidis, "Genocide of the Greeks"; Popov, *Culture, Ethnicity.*

28. Göçek, *Denial of Violence*; Neyzi, *Ben Kimim?*; Özyürek, *Politics of Public Memory in Turkey.*

29. Kirişci, "Migration and Turkey"; Zurcher, *Turkey.*

30. Yılmaz, *Bafra Ovası'nın Beşeri ve İktisadi Coğrafyası.*

31. Koylu and Doğan, "Birinci Dünya Savaşı Sırasında Osmanlı Devleti'nde Sıtma Mücedelesi," 209–215; "Sıtma Mücâdelesi Kânûnu."

32. Evered and Evered, "A Conquest of Rice."

33. Ahram, "Development, Counterinsurgency"; Evered, "Draining an Anatolian Desert"; Gratien, "The Ottoman Quagmire."

34. Yılmaz, *Bafra Ovası'nın Beşeri ve İktisadi Coğrafyası.*

35. "Bataklıkların Kurutulması ve Bundan Elde Edilecek Topraklar Hakkında Kanun."

36. Ayan, *Kızılırmak Deltasında Doğal Kaynak Kullanımı*; Yeniyürt et al., *Kızılırmak Deltası Sulak Alan Yönetim Planı.*

37. Republic of Turkey Ministry of Environment, *The Kızılırmak Delta.*

38. Özesmi, "Conservation Strategies."

39. For example, see Özgen and Taş, "Ramsar Alanı İçinde Yer Alan Cernek Gölü"; Republic of Turkey Ministry of Environment, *The Kızılırmak Delta.*

40. Republic of Turkey Ministry of Environment, *The Kızılırmak Delta*, 10.

41. Costanza, Farber, and Maxwell, "Valuation and Management of Wetland Ecosystems"; McAfee, "Selling Nature to Save It?"

42. Kasaba, *A Moveable Empire*, 21.

43. The word *longoz*, meaning "swamp forest" in Turkish, is a word of Greek origins. Its synonym, *su basan,* means literally "that which steps in water." Csató, Isaksson, and Jahani, *Linguistic Convergence and Areal Diffusion*, 338.

44. Ayan, *Kızılırmak Deltasında Doğal Kaynak Kullanımı*; Özesmi, "Ecological Economics of Harvesting Sharp-Pointed Rush."

45. Kohler, "Paul Errington."

46. Karaömerlioğlu, "Elite Perceptions of Land Reform."

47. Pigg, "Inventing Social Categories through Place."

48. Sirman, "State, Village, and Gender in Western Turkey."

Chapter 5

1. Çilingir, *Pontos Gerceği*; Hovannisian, *Armenian Pontus*; Göçek, *Denial of Violence*.

2. In conversations with sociologist Sezai Ozan Zeybek I learned of other conspiracies about the introduction of protected species in other regions of Turkey, particularly near mining sites. These conspiracies might be leveraged to counter environmental arguments against land or water expropriation for resource extraction. By contrast, the stories I heard in the Kızılırmak Delta pertained to the introduction of species that are not endemic or protected.

3. Govindrajan, "Monkey Business Macaque Translocation"; Subramaniam, "The Aliens Have Landed!"; Vitebsky, *The Reindeer People*.

4. Govindrajan, *Animal Intimacies*; Kirksey and Helmreich, "Emergence of Multispecies Ethnography"; Yates-Doerr, "Does Meat Come from Animals?"; Vitebsky, *The Reindeer People*.

5. Bird-Rose, "Flying Fox," 10; Haraway, *When Species Meet*.

6. Helmreich, *Alien Ocean*, 14.

7. Tambiah, "Animals Are Good to Think"; Bulmer, "Why Is the Cassowary Not a Bird?"; Mullin, "Mirrors and Windows."

8. Blanchette, "Herding Species"; Hansen, "Becoming Bovine"; Mikhail, "Unleashing the Beast"; Paxson, *The Life of Cheese*.

9. Cormier, *Kinship with Monkeys*; Govindrajan, "The Goat That Died for Family"; Charles, "Animals Just Love You as You Are."

10. Nadasdy, "The Gift in the Animal"; Kockelman, *The Chicken and the Quetzal*; Dave, "Witness."

11. On "making sense" of lived ecologies, see Scaramelli, "Making Sense of Water Quality."

12. Williams, "Dominant, Residual, and Emergent."

13. Doane, *Stealing Shining Rivers*; Walley, *Rough Waters*; West, *Conservation Is Our Government Now*.

14. Ayan, *Kızılırmak Deltasında Doğal Kaynak Kullanımı*; Ermetin, "Husbandry and Sustainability of Water Buffaloes in Turkey."

15. Karpat, "Social Effects of Farm Mechanization"; Kirişci, "Migration and Turkey."

16. UNDP GEF, *Conservation of Biodiversity.*

17. Yılmaz, Ertuğrul, and Wilson, "Domestic Livestock Resources of Turkey"; Selçuk, "Anadolu Mandalarının Doğal Yaşam Alanı 'Kızılırmak Deltası,'" 167.

18. Paxson, *The Life of Cheese.*

19. Bocci, "Tangles of Care"; Comaroff and Comaroff, "Naturing the Nation"; Helmreich, "How Scientists Think"; Subramaniam, "The Aliens Have Landed!"

20. Ayan, *Kızılırmak Deltasında Doğal Kaynak Kullanımı*; Özesmi, "Conservation Strategies"; Yılmaz, *Bafra Ovası'nın Beşeri ve İktisadi Coğrafyası*; Scaramelli, "Swamps into Wetlands"; Scaramelli, "The Wetland Is Disappearing."

21. Garcia-R and Trewick, "Dispersal and Speciation in Purple Swamphens."

22. Lopes et al., "Purple Swamphen."

23. BirdLife International, "Purple Swamphen."

24. Garcia-R and Trewick, "Dispersal and Speciation in Purple Swamphens."

25. Scaramelli, "The Wetland Is Disappearing."

26. Cronon, *Nature's Metropolis*; Haraway, "Teddy Bear Patriarchy"; Jacoby, *Crimes against Nature*; Powell, *Vanishing America*; Walley, *Rough Waters.*

27. Takacs, *The Idea of Biodiversity*; West, *Conservation Is Our Government Now.*

28. Hustings, and van Dijk, *Bird Census.*

29. Kirwan et al., *The Birds of Turkey*, 25.

30. Davidson and Kirwan, "Around the Region," 76–80.

31. Orbay, *Wetlands, the Source of Life.*

32. Demirbaş, "Kizilirmak Deltasi'nda Sazhorozu."

33. Demirbaş.

34. Del Hoyo, Elliott, and Sargatal, *Handbook of the Birds of the World.*

35. Mengüllüoğlu, Aktan, and Erdem, *Göksu Deltası.*

36. Ritvo, *Platypus and the Mermaid.*

37. Lopes et al., "Purple Swamphen."

38. Lowe, "Making the Monkey."

39. The sign, in Turkish, read "Kızılırmak Deltası'na Hoşgeldiniz."

40. I knew many of the contributors to the Facebook group for the delta in person. The page became the site for interesting debates and contestations on some of the most pivotal changes in the delta's built environment, ecology, and conservation policy. Here I am protecting my interlocutors' anonymity in this venue by not providing more context about their personal lives and work.

41. Çalışkan, "Explaining the End of Military Tutelary Regime."

42. "Orman Köylülerinin Kalkınmalarının Desteklenmesi."

43. Douglas, *Purity and Danger.*

44. Adams and McShane, *The Myth of Wild Africa*; Doane, *Stealing Shining Rivers*; Walley, *Rough Waters*; West, *Conservation Is Our Government Now.*

Conclusion

1. Turam, *Between Islam and the State.*

2. Hakyemez, *Turkey's Failed Peace Process.*

3. Eder and Özkul, "Precarious Lives and Syrian Refugees in Turkey." See also issue 288 of the *Middle East Research and Information Project (MERIP)* on "Confronting the New Turkey" (Fall 2018).

4. Bennett, *Vibrant Matter.*

5. McLean, "Black Goo."

6. Strang, *The Meaning of Water*; Orlove and Caton, "Water Sustainability."

7. Mitchell, "Can the Mosquito Speak?"

8. Barnes, *Cultivating the Nile.*

9. Helmreich, "Nature/Culture/Seawater."

10. Biersack, "Reimagining Political Ecology"; Escobar, "Constructing Nature"; Neumann, "Political Ecology of Wildlife Conservation"; Robbins, *Political Ecology*; Sivaramakrishnan, *Modern Forests.*

11. Raffles, *In Amazonia*; Vitebsky, *The Reindeer People*; Kirksey and Helmreich, "Emergence of Multispecies Ethnography"; Pandian, "Pastoral Power in the Postcolony"; Latour, *Reassembling the Social*; Bird-Rose, *Wild Dog Dreaming*; Van Dooren, *Flight Ways*; Yates-Doerr, "Does Meat Come from Animals?"; Blanchette, "Herding Species"; Govindrajan, *Animal Intimacies.*

12. Anand, "Pressure"; Farmer, "Cairo Ecologies"; Von Schnitzler, "Traveling Technologies"; Richardson and Weszkalnys, "Introduction: Resource Materialities"; Bruun Jensen, "Experimenting with Political Materials"; Choy and Zee, "Condition—Suspension"; Krause, "Making Space along the Kemi River."

Bibliography

Abu El-Haj, Nadia. *Facts on the Ground: Archaeological Practice and Territorial Self-Fashioning in Israeli Society*. Chicago: University of Chicago Press, 2008.

Adalet, Begüm. *Hotels and Highways: The Construction of Modernization Theory in Cold War Turkey*. Stanford, CA: Stanford University Press, 2018.

Adaman, Fikret, Bengi Akbulut, and Murat Arsel. *Neoliberal Turkey and Its Discontents: Economic Policy and the Environment under Erdogan*. London: I. B. Tauris, 2017.

Adaman, Fikret, and Murat Arsel. *Environmentalism in Turkey: Between Democracy and Development?* Hampshire, UK: Ashgate, 2005.

Adaman, Fikret, Serra Hakyemez, and Begüm Özkaynak. "The Political Ecology of a Ramsar Conservation Failure: The Case of Burdur Lake, Turkey." *Environment and Planning C: Politics and Space* 27, no. 5 (2009): 783–800.

Adams, Jonathan S., and Thomas O. McShane. *The Myth of Wild Africa: Conservation without Illusion*. Berkeley: University of California Press, 1996.

Agrawal, Arun. *Environmentality: Technologies of Government and Political Subjects*. Durham, NC: Duke University Press, 2005.

Ahram, Ariel I. "Development, Counterinsurgency, and the Destruction of the Iraqi Marshes." *International Journal of Middle East Studies* 47, no. 3 (2015): 447–466.

Akçam, Taner. *The Young Turks' Crime against Humanity: The Armenian Genocide and Ethnic Cleansing in the Ottoman Empire*. Princeton, NJ: Princeton University Press, 2012.

Aksu, Cemil, Sinan Erensü, and Erdem Evren, eds. *Sudan Sebepler: Türkiye'de Neoliberal Su-Enerji Politikaları ve Direnişleri*. Istanbul: İletişim Yayınları, 2016.

Aktay, Hatice. "İzmir Körfez Geçişi Projesi İptal Kararı İzmir Halkına Armağandır." *Sivil Sayfalar*, January 2, 2019. http://www.sivilsayfalar.org/2019/01/02/izmir-korfez-gecisi-projesi-iptal-karari-izmir-halkina-armagandir.

Alber, Erdmute, and Heike Drotbohm. *Anthropological Perspectives on Care: Work, Kinship, and the Life-Course*. New York: Palgrave Macmillan, 2015.

Alkan, Aybike. "Bilim ve Teknoloji Çalışmaları'nda Altyapılar: GAP'ın Kara Kutusu." *Toplum ve Bilim* 144 (2018): 105–124.

Alpbaz, Attila. "Homa Dalyanı'nın Su Ürünleri Fakültesine Tahsisi." Unpublished document. 2012.

Anand, Nikhil. *Hydraulic City: Water and the Infrastructures of Citizenship in Mumbai*. Durham, NC: Duke University Press, 2017.

———. "Pressure: The PoliTechnics of Water Supply in Mumbai." *Cultural Anthropology* 26, no. 4 (2011): 542–564.

Anand, Nikhil, Akhil Gupta, and Hannah Appel, eds. *The Promise of Infrastructure*. Durham, NC: Duke University Press, 2018.

Andersson, Tony. "Environmentalists with Guns: Conservation, Revolution, and Counterinsurgency in the Petén, Guatemala 1944–1996." PhD diss., New York University, 2018.

Arı, Yılmaz. "Visions of a Wetland: Linking Culture and Conservation at Lake Manyas, Turkey." PhD diss., University of Michigan, 2001.

Arsel, Murat, Bengi Akbulut, and Fikret Adaman. "Environmentalism of the Malcontent: Anatomy of an Anti–Coal Power Plant Struggle in Turkey." *Journal of Peasant Studies* 42, no. 2 (2015): 371–395.

Ash, Eric H. *The Draining of the Fens: Projectors, Popular Politics, and State Building in Early Modern England*. Baltimore: Johns Hopkins University Press, 2017.

Asif, Manan Ahmed, and Anand Vivek Taneja. "Introduction: Animals, Ethics, and Enchantment in South Asia and the Middle East." *Comparative Studies of South Asia, Africa and the Middle East* 35, no. 2 (2015): 200–203.

Atkinson-Willes, George. *Liquid Assets*. Gland, Switzerland: IUCN, 1965.

Ayan, Ali Kemal. *Kızılırmak Deltasında Doğal Kaynak Kullanımı*. Samsun, Turkey: Ondokuz Mayıs Üniversitesi, 2007.

Baker, Lauren, Samara Brock, Luisa Cortesi, Aysen Eren, Chris Hebdon, Francis Ludlow, Jeffrey Stoike, and Michael Dove. "Mainstreaming Morality: An Examination of Moral Ecologies as a Form of Resistance." *Journal for the Study of Religion, Nature and Culture* 11, no. 1 (2017): 23–55.

Ballestero, Andrea. "The Anthropology of Water." *Annual Review of Anthropology* 48, no. 1 (2019): 405–421.

———. "The Ethics of a Formula: Calculating a Financial–Humanitarian Price for Water." *American Ethnologist* 42, no. 2 (2015): 262–278.

———. *A Future History of Water*. Durham, NC: Duke University Press, 2019.

Barnes, Jessica. *Cultivating the Nile: The Everyday Politics of Water in Egypt*. Durham, NC: Duke University Press, 2014.

Barnes, Jessica, and Samer Alatout. "Water Worlds: Introduction to the Special Issue of Social Studies of Science." *Social Studies of Science* 42, no. 4 (2012): 483–488.

Barnes, Jessica, and Mariam Taher. "Care and Conveyance: Buying Baladi Bread in Cairo." *Cultural Anthropology* 34, no. 3 (2019): 417–443.

Bartu Candan, Ayfer. "Remembering a 9000 Years Old Site: Present-ing Çatalhöyük." In *Politics of Public Memory: Production and Consumption of the Past in Turkey*, edited by Esra Özyürek, 70–94. Syracuse, NY: Syracuse University Press, 2007.

Bartu Candan, Ayfer, and Biray Kolluoğlu. "Emerging Spaces of Neoliberalism: A Gated Town and a Public Housing Project in İstanbul." *New Perspectives on Turkey* 39 (2008): 5–46.

"Bataklıkların Kurutulması ve Bundan Elde Edilecek Topraklar Hakkında Kanun." *Resmî Gazete* 7413 (January 18, 1950): 17649–17650. https://www.resmigazete.gov.tr/arsiv/7413.pdf.

Bayram, Ali. "İzmir Kuş Cenneti'nde Başı Boş Atların Sayısı Artıyor." *Hürriyet Ege*, October 30, 2013. https://www.haberler.com/izmir-kus-cenneti-nde-basi-bos-atlarin-sayisi-5235207-haberi/.

Beaman, Mark, Richard F. Porter, and Alan Vittery, eds. *Bird Report 1970–1973*. London: Ornithological Society of Turkey, 1975.

Bellacasa, Maria Puig de la. "Making Time for Soil: Technoscientific Futurity and the Pace of Care." *Social Studies of Science* 45, no. 5 (2015): 691–716.

Belon, Pierre. *L'histoire de la nature des oyseaux, avec leurs descriptions, & naïfs portraicts retirez du naturel*. Paris: Chez Guillaume Cauellat, deuant le College de Cambray, à l'enseigne de la Pouïle Grasse: [Imprime par Benoist Preuost, demeurant en la rue Fremontel, prés le cloz Bruneau, à l'enseigne de l'Estoille d'Or], 1555.

Bennett, Jane. *Vibrant Matter: A Political Ecology of Things*. Durham, NC: Duke University Press, 2009.

Berry, John. *International Wildfowl Inquiry*. Vol. 1, *Factors Affecting the General Status of Wild Geese and Wild Duck*. Cambridge: Cambridge University Press, 1941.

———. *International Wildfowl Inquiry*. Vol. 2, *The Status and Distribution of Wild Geese and Wild Duck in Scotland*. Cambridge: Cambridge University Press, 1939.

Bhattacharyya, Debjani. *Empire and Ecology in the Bengal Delta: The Making of Calcutta*. Cambridge: Cambridge University Press, 2018.

Biersack, Aletta. "Reimagining Political Ecology: Culture/Power/History/Nature." In *Reimagining Political Ecology*, edited by James B. Greenberg and Aletta Biersack, 3–40. Durham, NC: Duke University Press, 2006.

Biggs, David. *Quagmire: Nation-Building and Nature in the Mekong Delta*. Seattle: University of Washington Press, 2011.

BirdLife International. "Purple Swamphen." *The IUCN Red List of Threatened Species*. 2016. https://www.iucnredlist.org/species/22692792/155531172.

Bird-Rose, Deborah. "Flying Fox: Kin, Keystone, Kontaminant." *Manoa* 22, no. 2 (2010): 175–190.

———. *Wild Dog Dreaming: Love and Extinction*. Charlottesville: University of Virginia Press, 2011.

Björkman, Lisa. *Pipe Politics, Contested Waters: Embedded Infrastructures of Millennial Mumbai*. Durham, NC: Duke University Press, 2015.

Blackbourn, David. *The Conquest of Nature: Water, Landscape, and the Making of Modern Germany*. New York: W. W. Norton, 2007.

Blanchette, Alex. "Herding Species: Biosecurity, Posthuman Labor, and the American Industrial Pig." *Cultural Anthropology* 30, no. 4 (2015): 640–669.

Bocci, Paolo. "Tangles of Care: Killing Goats to Save Tortoises on the Galápagos Islands." *Cultural Anthropology* 32, no. 3 (2017): 424–449.

Bozdogan, Sibel. *Modernism and Nation Building: Turkish Architectural Culture in the Early Republic.* Seattle: University of Washington Press, 2001.

Brooks, Duncan J. *Sandgrouse.* Vol. 11. Sandy, Bedfordshire, UK: Ornithological Society of the Middle East, 1989.

Bruun Jensen, Casper. "Amphibious Worlds: Environments, Infrastructures, Ontologies." *Engaging Science, Technology, and Society* 3 (2017): 224–232.

———. "Experimenting with Political Materials: Environmental Infrastructures and Ontological Transformations." *Scandinavian Journal of Social Theory* 16, no. 1 (2015): 17–30.

———. "The *Umwelten* of Infrastructure: A Stroll along (and inside) Phnom Penh's Sewage Pipes." *Zinbun* 57 (2017): 147–159.

Bruyninckx, Joeri. *Listening in the Field: Recording and the Science of Birdsong.* Cambridge, MA: MIT Press, 2018.

Bubandt, Nils, and Anna Tsing. "Feral Dynamics of Post-industrial Ruin: An Introduction." *Journal of Ethnobiology* 38, no. 1 (2018): 1–7.

Buch, Elana D. "Anthropology of Aging and Care." *Annual Review of Anthropology* 44, no. 1 (2015): 277–293.

Bulmer, Ralph. "Why Is the Cassowary Not a Bird? A Problem of Zoological Taxonomy among the Karam of the New Guinea Highlands." *Man* 2, no. 1 (1967): 5–25.

Çalışkan, Koray. "Explaining the End of Military Tutelary Regime and the July 15 Coup Attempt in Turkey." *Journal of Cultural Economy* 10, no. 1 (2017): 97–111.

Campbell, Ben. "Moral Ecologies of Subsistence and Labour in a Migration-Affected Community of Nepal." *Journal of the Royal Anthropological Institute* 24, no. 1 (2018): 151–165.

Carp, Erik. *Directory of Wetlands of International Importance in the Western Palearctic.* Gland, Switzerland: IUCN, UNEP, WWF, 1980.

———. *Final Act and Summary Record: International Conference on the Conservation of Wetlands and Waterfowl, Ramsar Iran 1971.* Slimbridge, UK: International Waterfowl Research Bureau, 1972.

Carroll, Patrick. "Water and Technoscientific State Formation in California." *Social Studies of Science* 42, no. 4 (2012): 489–516.

Carse, Ashley. *Beyond the Big Ditch: Politics, Ecology, and Infrastructure at the Panama Canal.* Cambridge, MA: MIT Press, 2014.

———. "Infrastructure—How a Humble French Engineering Term Shaped the Modern World." In *Infrastructures and Social Complexity*, edited by Penny Harvey and Atsuro Morita, 27–39. New York: Routledge, 2017.

———. "Nature as Infrastructure: Making and Managing the Panama Canal Watershed." *Social Studies of Science* 42, no. 4 (2012): 539–563.

Cassidy, Rebecca, and Molly Mullin. *Where the Wild Things Are Now: Domestication Reconsidered*. Oxford: Berg, 2007.

Cézilly, Frank, and Tobias Salathé. "Luc Hoffmann (1923–2016)." *Ibis* 159, no. 2 (2017): 472–474.

Charles, Nickie. "'Animals Just Love You as You Are': Experiencing Kinship across the Species Barrier." *Sociology* 4 (2014): 715–730.

Choy, Timothy. *Ecologies of Comparison: An Ethnography of Endangerment in Hong Kong*. Durham, NC: Duke University Press, 2011.

Choy, Timothy, and Jerry Zee. "Condition—Suspension." *Cultural Anthropology* 30, no. 2 (2015): 210–223.

Clark, Bruce. *Twice a Stranger: The Mass Expulsions That Forged Modern Greece and Turkey*. Cambridge, MA: Harvard University Press, 2006.

Coleman, Leo. *A Moral Technology: Electrification as Political Ritual in New Delhi*. Ithaca, NY: Cornell University Press, 2017.

Comaroff, Jean, and John L. Comaroff. "Naturing the Nation: Aliens, Apocalypse, and the Postcolonial State." *Journal of Southern African Studies* 27, no. 3 (2001): 627–651.

Cons, Jason. "Staging Climate Security: Resilience and Heterodystopia in the Bangladesh Borderlands." *Cultural Anthropology* 33, no. 2 (2018): 266–294.

Cormier, Loretta A. *Kinship with Monkeys: The Guajá Foragers of Eastern Amazonia*. New York: Columbia University Press, 2003.

Costanza, Robert, Ralph d'Arge, Rudolf de Groot, Stephen Farber, Monica Grasso, Bruce Hannon, Karin Limburg, et al. "The Value of the World's Ecosystem Services and Natural Capital." *Nature* 387 (1997): 253–260.

Costanza, Robert, Stephen C. Farber, and Judith Maxwell. "Valuation and Management of Wetland Ecosystems." *Ecological Economics* 1, no. 4 (1989): 335–361.

Cronon, William. *Nature's Metropolis: Chicago and the Great West*. New York: W. W. Norton, 1992.

———. "The Trouble with Wilderness; or, Getting Back to the Wrong Nature." In *Uncommon Ground: Rethinking the Human Place in Nature*, edited by William Cronon, 60–90. New York: W. W. Norton, 1995.

Csató, Éva Ágnes, Bo Isaksson, and Carina Jahani. *Linguistic Convergence and Areal Diffusion: Case Studies from Iranian, Semitic and Turkic*. London: Psychology Press, 2005.

Çilingir, Tamer. *Pontos Gerçeği: 1914–1923 Yılları Arasında Karadeniz'de Yaşananlar*. Istanbul: Belge Yayınevi, 2016.

Da Cunha, Dilip. *The Invention of Rivers: Alexander's Eye and Ganga's Descent*. Philadelphia: University of Pennsylvania Press, 2019.

Dalyan, Can. "Latent Lives: Genebanking and the Politics of Conservation in Turkey." PhD diss., Cornell University, 2018.

Darkot, Besim. "Ege Haliçlerinin Mense ve Tekamülü." In *Coğrafi Araştırmalar*, by Besim Darkot, 29–52, 93–95. Istanbul: Edebiyat Fakültesi Coğrafia Basımevi, 1938.

Datta, Ayona. "Gendered Nature and Urban Culture: The Dialectics of Gated Developments in Izmir, Turkey." *International Journal of Urban and Regional Research* 38, no. 4 (2014): 1363–1383.

Dave, Naisargi N. "Witness: Humans, Animals, and the Politics of Becoming." *Cultural Anthropology* 29, no. 3 (2014): 433–456.

Davidson, Pete, and Guy M. Kirwan. "Around the Region." *Sandgrouse* 18, no. 1 (1996): 76–80.

Davies, Jon, and Gordon Claridge. *Wetland Benefits: The Potential for Wetlands to Support and Maintain Development*. Kuala Lumpur: Asian Wetland Bureau, International Waterfowl and Wetlands Research Bureau, Wetlands for the Americas, 1993.

Davis, Diana K., ed. *Environmental Imaginaries of the Middle East and North Africa*. Athens: Ohio University Press, 2013.

De Klemm, Cyril, and Isabelle Créteaux. *The Legal Development of the Ramsar Convention on Wetlands of International Importance Especially as Waterfowl Habitat (2 February 1971)*. Gland, Switzerland: Ramsar Convention Bureau, 1993.

Del Hoyo, Joseph, Andrew Elliott, and Jordi Sargatal, eds. *Handbook of the Birds of the World*. Vol. 3. Barcelona: Lynx Editions, 1996.

Demirbaş, Gökçen. "Kizilirmak Deltasi'nda Sazhorozu Populasyonunun Saptanmasi." PhD diss., Ondokuz Mayıs Üniversitesi, 2007.

Doane, Molly. *Stealing Shining Rivers: Agrarian Conflict, Market Logic, and Conservation in a Mexican Forest*. Tucson: University of Arizona Press, 2012.

Doğa Derneği. "Flamingo Bölgesine Otoban." *MAGMA*, June 30, 2017. http://www.magmadergisi.com/direndoga/flamingo-bolgesine-otoban.

———. "Gediz'in Vazgeçilmesi: Atlar." *Atlas Dergisi*, May 30, 2012. http://atlasdergisi.net/gedizin-vazgecilmezi-atlar/3035n.aspx [site discontinued].

Doğal Hayatı Koruma Derneği (DHKD). *Göksu Delta*. Ankara: DHKD, 1995.

———. *Towards Integrated Management in the Göksu Delta, a Protected Special Area in Turkey*. Ankara: DHKD, 1992.

Doğer, Ersin. *Ilk Insanlardan Yunan Işgaline Kadar Menemen Ya Da Tarhaniyat Tarihi*. Izmir: Sergi Yayınevi, 1997.

Dolbee, Sam. "The Desert at the End of Empire: An Environmental History of the Armenian Genocide." *Past & Present* 247, no 1 (2020): 197–233.

Dole, Christopher. *Healing Secular Life: Loss and Devotion in Modern Turkey*. Philadelphia: University of Pennsylvania Press, 2012.

Doughty, Robin W. *Feather Fashions and Bird Preservation: A Study in Nature Protection*. Berkeley: University of California Press, 1975.

Douglas, Mary. *Purity and Danger: An Analysis of Concepts of Pollution and Taboo*. London: Routledge, 1966.

Dove, Michael R., and Daniel M. Kammen. "The Epistemology of Sustainable Resource Use: Managing Forest Products, Swiddens, and High-Yielding Variety Crops." *Human Organization* 56, no. 1 (1997): 91–101.

Dresser, Henry E. "On a Collection of Birds from Erzeroom." *Ibis* 33, no. 3 (1891): 364–370.

Dunlap, Thomas R. *In the Field, among the Feathered: A History of Birders and Their Guides*. London: Oxford University Press, 2011.

"Dünyanın 3. Büyük Flamingo Adası." *Haberler.Com*, March 5, 2012. https://www.haberler.com/dunyanin-3-buyuk-flamingo-adasi-3418969-haberi/?utm_source=facebook&utm_campaign=tavsiye_et&utm_medium=detay.

Edelman, Marc. "Bringing the Moral Economy Back in . . . to the Study of 21st-Century Transnational Peasant Movements." *American Anthropologist* 107, no. 3 (2005): 331–345.

Eder, Mine, and Derya Özkul. "Editors' Introduction: Precarious Lives and Syrian Refugees in Turkey." *New Perspectives on Turkey* 54 (2016): 1–8.

Egemen, Medih. *Türkiyede Tuzculuk ve Çamaltı Tuzlası*. Istanbul: Tekel İnstitüleri Yayınları, 1946.

Ekmekcioglu, Lerna. *Recovering Armenia: The Limits of Belonging in Post-genocide Turkey*. Stanford, CA: Stanford University Press, 2016.

Elyachar, Julia. "Phatic Labor, Infrastructure, and the Question of Empowerment in Cairo." *American Ethnologist* 37, no. 3 (2010): 452–464.

EPFT. *The Wetlands of Turkey*. Translated by Virginia Taylor-Saçlıoğlu. Ankara: Türkiye Çevre Sorunları Vakfı, 1989.

Erdem, Osman. *Sulak Alanların Önemi ve Türkiye'nin "A" Sınıfı Sulak Alanları*. Ankara: T. C. Çevre Bakanlığı, 1994.

———. *Turkey's Bird Paradises*. Green Series 5. Ankara: Republic of Turkey Ministry of Environment and Forestry General Directorate of Environmental Protection, 1995.

Erensü, Sinan. "Abundance and Scarcity amidst the Crisis of 'Modern Water': The Changing Water-Energy Nexus in Turkey." In *Contemporary Water Governance in the Global South: Scarcity, Marketization and Participation*, edited by Leila M. Harris, Jacqueline A. Goldin, and Christopher Sneddon, 61–78. New York: Routledge, 2015.

———. "Fragile Energy: Power, Nature, and the Politics of Infrastructure in the New Turkey." PhD diss., University of Minnesota, 2016.

———. "Powering Neoliberalization: Energy and Politics in the Making of a New Turkey." *Energy Research & Social Science* 41 (2018): 148–157.

Erensü, Sinan, and Ayça Alemdaroğlu. "Dialectics of Reform and Repression: Unpacking Turkey's Authoritarian 'Turn.'" *Review of Middle East Studies* 52, no. 1 (2018): 16–28.

Erensü, Sinan, and Ozan Karaman. "The Work of a Few Trees: Gezi, Politics and Space." *International Journal of Urban and Regional Research* 41, no. 1 (2016): 19–36.

Erman, Deniz Onur. "Bird Houses in Turkish Culture and Contemporary Applications." *Procedia—Social and Behavioral Sciences* 122 (2014): 306–311.

Ermetin, Orhan. "Husbandry and Sustainability of Water Buffaloes in Turkey." *Turkish Journal of Agriculture, Food Science, and Technology* 5, no. 12 (2017): 1673–1682.

Ernoul, Lisa. "Entre Camargue et Delta de Gediz: Réflexions sur les transferts de modèles de gestion intégrée des zones cotières." PhD diss., Université d'Aix-Marseille, 2014.

Ernoul, Lisa, and Angela Wardell-Johnson. "Governance in Integrated Coastal Zone Management: A Social Networks Analysis of Cross-scale Collaboration." *Environmental Conservation* 40, no. 3 (2013): 231–240.

Erol, Emre. *The Ottoman Crisis in Western Anatolia: Turkey's Belle Epoque and the Transition to a Modern Nation State.* London: I. B. Taurus, 2016.

Errington, Frederick, and Deborah Gewertz. "Managing an Endangered Species: Palliative Care for the Pallid Sturgeon." *American Ethnologist* 45, no. 2 (2018): 186–200.

Errington, Paul Lester. *Of Men and Marshes.* New York: Macmillan, 1957.

Ertan, Asaf, Aygün Kılıç, and Max Kasparek. *Türkiy'nin Önemli Kuş Alanları.* Istanbul: Doğal Hayatı Koruma Derneği, 1989.

Escobar, Arturo. "Constructing Nature, Elements for a Poststructural Political Ecology." In *Liberation Ecologies, Environment, Development, Social Movements,* edited by Richard Peet and Michael Watts, 46–68. New York: Routledge, 1996.

Evered, Kyle T. "Draining an Anatolian Desert: Overcoming Water, Wetlands, and Malaria in Early Republican Ankara." *Cultural Geographies* 21, no. 3 (2013): 1–22.

Evered, Kyle T., and Emine Ö. Evered. "A Conquest of Rice: Agricultural Expansion, Impoverishment, and Malaria in Turkey." *Historia Agraria* 68 (2016): 103–136.

———. "Governing Population, Public Health, and Malaria in the Early Turkish Republic." *Journal of Historical Geography* 37 (2011): 470–482.

Evren, Erdem. "The Rise and Decline of an Anti-dam Campaign: Yusufeli Dam Project and the Temporal Politics of Development." *Water History* 6, no. 4 (2014): 405–419.

Farmer, Tessa. "Cairo Ecologies: Water in Social and Material Cycles." PhD diss., University of Texas at Austin, 2014.

———. "Willing to Pay: Competing Paradigms about Resistance to Paying for Water Services in Cairo, Egypt." *Middle East Law and Governance* 9 (2017): 3–19.

Faroqhi, Suraiya, ed. *Animals and People in the Ottoman Empire.* Istanbul: Eren, 2010.

Finlayson, C. Max, Nick Davidson, Dave Pritchard, G. Randy Milton, and Heather MacKay. "The Ramsar Convention and Ecosystem-Based Approaches to the Wise Use and Sustainable Development of Wetlands." *Journal of International Wildlife Law & Policy* 14, no. 3–4 (2011): 176–198.

Fırat, Bilge. "The Most Eastern of the West, the Most Western of the East: Energy-Transport Infrastructures and Regional Politics of the Periphery in Turkey." *Economic Anthropology* 3 (2016): 81–93. [she moved this so does this indicate something about pronunciation of dotless i]

First Annual Report of Severn Wildfowl Trust: 1948. Slimbridge, UK: Severn Wildfowl Trust, 1948.

Fischer, Michael M. J. "Emergent Forms of Life: Anthropologies of Late or Postmodernities." *Annual Review of Anthropology* 28, no. 1 (1999): 455–478.

Gagné, Karine. *Caring for Glaciers: Land, Animals, and Humanity in the Himalayas.* Seattle: University of Washington Press, 2019.

Garcia-R, Juan C., and Steve A. Trewick. "Dispersal and Speciation in Purple Swamphens (Rallidae: Porphyrio)." *The Auk* 132, no. 1 (2014): 140–155.

"George Atkinson-Willes." *The Telegraph*, January 3, 2003. https://www.telegraph.co.uk/news/obituaries/1417706/George-Atkinson-Willes.html.

Gewertz, Deborah, and Frederick Errington. "Doing Good and Doing Well: Prairie Wetlands, Private Property, and the Public Trust." *American Anthropologist* 114, no. 1 (2015): 17–31.

———. "Pheasant Capitalism: Auditing South Dakota's State Bird." *American Ethnologist* 42, no. 3 (2015): 399–414.

Geylan, Etem. "Sulak Alanların Kuruduğu Algısı Gerçeği Yansıtmıyor." *Haberler.* February 1, 2017. https://www.haberler.com/sulak-alanlarin-kurudugu-algisi-gercegi-9218926-haberi/.

Giblett, Rodney James. *Postmodern Wetlands: Culture, History, Ecology.* Edinburgh: Edinburgh University Press, 1996.

Göçek, Fatma Müge. *Denial of Violence: Ottoman Past, Turkish Present and Collective Violence against the Armenians, 1789–2009.* Oxford: Oxford University Press, 2015.

Goffman, Daniel. *Izmir and the Levantine World 1550–1650.* Seattle: University of Washington Press, 1990.

———. *The Ottoman Empire and Early Modern Europe.* Cambridge: Cambridge University Press, 2002.

Gökay, Bülent, and Tunç Aybak. "Identity, Race and Nationalism in Turkey: Introduction to the Special Issue." *Journal of Balkan and Near Eastern Studies* 18, no. 2 (2016): 107–110.

Govindrajan, Radhika. *Animal Intimacies: Interspecies Relatedness in India's Central Himalayas.* Chicago: University of Chicago Press, 2018.

———. "'The Goat That Died for Family': Animal Sacrifice and Interspecies Kinship in India's Central Himalayas." *American Ethnologist* 42, no. 3 (2015): 504–519.

———. "Monkey Business Macaque Translocation and the Politics of Belonging in India's Central Himalayas." *Comparative Studies of South Asia, Africa and the Middle East* 35, no. 2 (2015): 246–262.

Grasseni, Cristina. "Skilled Vision: An Apprenticeship in Breeding Aesthetics." *Social Anthropology* 12, no. 1 (2004): 41–55.

Gratien, Chris. "The Ottoman Quagmire: Malaria, Swamps, and Settlement in the Late Ottoman Mediterranean." *International Journal of Middle East Studies* 49 (2017): 583–604.

Groot, Rudolf de, Mishka Stuip, and Max Finlayson. *Guidance for Valuing the Benefits Derived from Wetland Ecosystem Services.* CBD Technical Series No. 27. Gland, Switzerland: Ramsar Convention Secretariat, 2006.

Gruppuso, Paolo. "Edenic Views in Wetland Conservation: Nature and Agriculture in the Fogliano Area, Italy." *Conservation and Society* 16, no. 4 (2018): 397–408.

Guarasci, Brigitte. "The National Park: Restoring the Marshes in Wartime Iraq." *Arab Studies Journal* 23, no. 1 (2015): 128–153.

Güçlü, Kamuran, and Faris Karahan. "A Review: The History of Conservation Programs and Development of the National Parks Concept in Turkey." *Biodiversity & Conservation* 13, no. 7 (2004): 1373–1390.

Gültekin, Turan. "Flamingolar Yaşasın Diye Her Cuma Eylem." *Hürriyet*, October 4, 2013. https://www.hurriyet.com.tr/ege/flamingolar-yasasin-diye-her-cuma-eylem-24844739.

Gündoğdu, Cihangir. "The State and the Stray Dogs in Late Ottoman Istanbul: From Unruly Subjects to Servile Friends." *Middle Eastern Studies* 54, no. 4 (2018): 555–574.

Gürpinar, Tansu. "Doğal Çevre Ve Fotoğraf: Sulak Alanlar." *Fotoritim* 20, August 5, 2008. http://www.arsivfotoritim.com/page/211/ [site discontinued].

Hagen, Joel B. *An Entangled Bank: The Origins of Ecosystem Ecology*. New Brunswick, NJ: Rutgers University Press, 1992.

Hakyemez, Serra. *Turkey's Failed Peace Process with the Kurds: A Different Explanation*. Middle East Brief. Waltham, MA: Brandeis University Crown Center for Middle East Studies, 2017.

Hansen, Paul. "Becoming Bovine: Mechanics and Metamorphosis in Hokkaido's Animal-Human-Machine." *Journal of Rural Studies* 33 (2014): 119–130.

Haraway, Donna. "Teddy Bear Patriarchy: Taxidermy in the Garden of Eden, New York City, 1908–1936." *Social Text* 11 (1984): 20–64.

———. *When Species Meet*. Minneapolis: University of Minnesota Press, 2008.

Harris, Leila M. "Contested Sustainabilities: Assessing Narratives of Environmental Change in Southeastern Turkey." *Local Environment* 14, no. 8 (2009): 699–720.

———. "Irrigation, Gender, and Social Geographies of the Changing Waterscapes of Southeastern Anatolia." *Environment and Planning D: Society and Space* 24, no. 2 (2006): 187–213.

Hart, Kimberly. *And Then We Work for God: Rural Sunni Islam in Western Turkey*. Stanford, CA: Stanford University Press, 2013.

———. "The Suburbanization of Rural Life in an Arid and Rocky Village in Western Turkey." *Journal of Arid Environments* 149 (2017): 1–7.

Harvey, Penny, and Hannah Knox. "The Enchantments of Infrastructure." *Mobilities* 7, no. 4 (2012): 521–536.

———. *Roads: An Anthropology of Infrastructure and Expertise*. Ithaca, NY: Cornell University Press, 2015.

Hathaway, Michael J. *Environmental Winds: Making the Global in Southwest China*. Berkeley: University of California Press, 2013.

Hecht, Gabrielle. *Being Nuclear: Africans and the Global Uranium Trade*. Cambridge, MA: MIT Press, 2014.

"Hedef 2023." *AK Parti*. https://web.archive.org/web/20200204005852/https://www.akparti.org.tr/hedef-2023/.

Helmreich, Stefan. *Alien Ocean: Anthropological Voyages in Microbial Seas.* Berkeley: University of California Press, 2009.

———. "How Scientists Think; about 'Natives,' for Example: A Problem of Taxonomy among Biologists of Alien Species in Hawaii." *Journal of the Royal Anthropological Institute* 11, no. 1 (2005): 107–128.

———. "Nature/Culture/Seawater." *American Anthropologist* 113, no. 1 (2011): 132–144.

Hetherington, Kregg, ed. *Infrastructure, Environment, and Life in the Anthropocene.* Durham, NC: Duke University Press, 2019.

———. "Keywords of the Anthropocene." In *Infrastructure, Environment, and Life in the Anthropocene,* edited by Kregg Hetherington, 1–16. Durham, NC: Duke University Press, 2019.

Hoffmann, Luc. "Research in Wetland Reserves." In *Proceedings of the XVI International Congress of Zoology,* edited by John A. Moore, 3:400–405. Washington, DC: XVI International Congress of Zoology, 1963.

Holleman, Hannah. *Dust Bowls of Empire: Imperialism, Environmental Politics, and the Injustice of "Green" Capitalism.* New Haven, CT: Yale University Press, 2018.

Hovannisian, Richard G. *Armenian Pontus: The Trebizond-Black Sea Communities.* Costa Mesa, CA: Mazda Publishers, 2009.

Howe, Cymene, Jessica Lockrem, Hannah Appel, Edward Hackett, Dominic Boyer, Randal Hall, Matthew Schneider-Mayerson, et al. "Paradoxical Infrastructures: Ruins, Retrofit, and Risk." *Science, Technology, & Human Values* 41, no. 3 (2016): 547–565.

Hughes, David McDermott. "Water as Boundary: National Parks, Rivers, and the Politics of Demarcation in Chinanimani, Zimbabwe." In *Reflections on Water: New Approaches to Transboundary Conflicts and Cooperation,* edited by Joachim Blatter and Helen Ingram, 267–294. Cambridge, MA: MIT Press, 2011.

Hughes, Thomas P. "The Evolution of Large Technological Systems." In *The Social Construction of Technological Systems: New Directions in the Sociology and History of Technology,* edited by Thomas P. Hughes, Wiebe E. Bijker, and Pinch Trevor, 51–82. Cambridge, MA: MIT Press, 1987.

Husain, Faisal. "Changes in the Euphrates River: Ecology and Politics in a Rural Ottoman Periphery, 1687–1702." *Journal of Interdisciplinary History* 47, no. 1 (2016): 1–25.

———. "In the Bellies of the Marshes: Water and Power in the Countryside of Ottoman Baghdad." *Environmental History* 19, no. 4 (2014): 638–664.

Hustings, Fred, and Klaas van Dijk, eds. *Bird Census in the Kızılırmak Delta, Turkey, in Spring 1992.* Zeist, Netherlands: WIWO, 1994.

Ignatow, Gabriel. "Economic Dependence and Environmental Attitudes in Turkey." *Environmental Politics* 14, no. 5 (2005): 648–666.

Igoe, Jim, and Dan Brockington. "Neoliberal Conservation: A Brief Introduction." *Conservation and Society* 5, no. 4 (2007): 432–449.

Ilcan, Suzan, and Lynne Phillips. "Developmentalities and Calculative Practices: The Millennium Development Goals." *Antipode* 42, no. 4 (2010): 844–874.

"In Memoriam: George Atkinson-Willes." *Ramsar*, December 12, 2002. https://www.ramsar.org/fr/node/38459.

IUCN, IUBP, and IWRB. *Proceedings of a Technical Meeting on Wetland Conservation, Ankara-Bursa-Istanbul, 9 to 16 October 1967*. Morges, Switzerland: IUCN Publications, 1968.

"İzmir'in Gediz Deltası UNESCO Dünya Doğa Mirası Ilan Edilsin!" *Gediz Mirasimizdir*, 2019. https://gedizmirasimizdir.org.

"İzmir Körfezi'ni Bekleyen Büyük Tehlike." *Izmir Büyükşehir Belediyesi*, November 26, 2014. https://www.izmir.bel.tr/tr/Haberler/izmir-korfezini-bekleyen-buyuk-tehlike/11902/156.

Jacobs, Nancy J. *Birders of Africa: History of a Network*. New Haven, CT: Yale University Press, 2016.

Jacoby, Karl. *Crimes against Nature: Squatters, Poachers, Thieves, and the Hidden History of American Conservation*. Berkeley: University of California Press, 2014.

Jax, Kurt, David N. Barton, Kai M. A. Chan, Rudolf de Groot, Ulrike Doyle, Uta Eser, Christoph Görg, et al. "Ecosystem Services and Ethics." *Ecological Economics* 93 (2013): 260–268.

Jensen, Casper Bruun. "The *Umwelten* of Infrastructure: A Stroll along (and inside) Phnom Penh's Sewage Pipes." *Zinbun* 57 (2017): 147–159.

Kadioğlu, Ayşe. "The Paradox of Turkish Nationalism and the Construction of Official Identity." *Middle Eastern Studies* 32, no. 2 (1996): 177–193.

Kaplan, Adnan, and Şerif Hepcan. "An Examination of Ecological Risk Assessment at Landscape Scale and the Management Plan." In *Decision Support for Natural Disasters and Intentional Threats to Water Security*, edited by T. H. Illangasekare, 237–251. Dordrecht, Netherlands: Springer Science, 2009.

Karademir Erol, Gamze. "Körfezin Son Dalyanı Homa Koruma Altında." *Egeden* 4, no. 14 (2012): 10–13.

Karadeniz, Nilgül, Alpay Tırıl, and Emel Baylan. "Wetland Management in Turkey: Problems, Achievements and Perspectives." *African Journal of Agricultural Research* 4, no. 11 (2009): 1106–1119.

Karakoyun, Umut. "İzmir Körfez Geçişi'ne Durdurma: Flamingolara Iyi Haber." *Hürriyet*, August 13, 2018. http://www.hurriyet.com.tr/gundem/izmir-korfez-gecisi-ne-durdurma-flamingolara-iyi-haber-40927639.

Karaömerlioğlu, M. Asim. "Elite Perceptions of Land Reform in Early Republican Turkey." *Journal of Peasant Studies* 27, no. 3 (2000): 115–141.

"Kararname Ekleyen Kamu Müteahhitleri, Temsili Cenaze Namazı Kıldı." *Hürriyet*, June 6, 2018. http://www.hurriyet.com.tr/yerel-haberler/adana/cukurova/karar-name-bekleyen-kamu-muteahhitleri-temsili-cenaze-namazi-kildi-41010051.

Karpat, Kemal H. "Social Effects of Farm Mechanization in Turkish Villages." *Social Research* 27, no. 1 (1960): 83–103.

Kasaba, Reşat. *A Moveable Empire: Ottoman Nomads, Migrants, and Refugees*. Seattle: University of Washington Press, 2009.

Kasdoğan, Duygu. "Potentiating Algae, Modernizing Bioeconomies: Algal Biofuels, Bioenergy Economies, and Built Ecologies in the United States and Turkey." PhD diss., York University, 2017.

Kayaalp, Ebru. *Remaking Politics, Markets, and Citizens in Turkey: Governing through Smoke*. London: Bloomsbury Publishing, 2014.

Keck, Margaret E., and Kathryn Sikkink. *Activists beyond Borders: Advocacy Networks in International Politics*. Ithaca, NY: Cornell University Press, 1998.

Kim, Eleana. "Toward an Anthropology of Landmines: Rogue Infrastructure and Military Waste in the Korean DMZ." *Cultural Anthropology* 31, no. 2 (2015): 162–187.

Kirişci, Kemal. "Migration and Turkey: The Dynamics of State, Society and Politics." In *The Cambridge History of Turkey*, vol. 4, *Turkey in the Modern World*, edited by Reşat Kasaba, 266–300. Cambridge: Cambridge University Press, 2008.

Kirksey, Eben. *Emergent Ecologies*. Durham, NC: Duke University Press, 2015.

Kirksey, Eben S., and Stefan Helmreich. "The Emergence of Multispecies Ethnography." *Cultural Anthropology* 24, no. 5 (2010): 545–575.

Kirwan, Guy, Barbaros Demirci, Hilary Welch, Kerem Boyla, Metehan Özen, Peter Castell, and Tim Marlow. *The Birds of Turkey*. New York: Bloomsbury Publishing, 2010.

Kitson, Alan R., and Richard F. Porter, eds. *Bulletin No. 5*. Sandy, Bedfordshire, UK: Ornithological Society of Turkey, 1970.

"Kızılırmak Deltası, Gediz Deltası, Akyatan Lagünü ve Uluabat Gölü Koruma Alanı Ilan Edilecek." *Cumhuriyet*, February 3, 1998. https://www.cumhuriyetarsivi.com/katalog/192/sayfa/1998/2/3/18.xhtml.

Kızılkaya, Zafer. "Homa Dalyanı'nın Yeni Yüzü." *Altas Dergisi* 223 (2012): 28.

Knudsen, Ståle. "Protests against Energy Projects in Turkey: Environmental Activism above Politics?" *British Journal of Middle Eastern Studies* 3 (2016): 302–323.

Kockelman, Paul. *The Chicken and the Quetzal: Incommensurate Ontologies and Portable Values in Guatemala's Cloud Forest*. Durham, NC: Duke University Press, 2016.

Kohler, Robert E. *Landscapes and Labscapes: Exploring the Lab-Field Border in Biology*. Chicago: University of Chicago Press, 2002.

———. "Paul Errington, Aldo Leopold, and Wildlife Ecology: Residential Science." *Historical Studies in the Natural Sciences* 41, no. 1 (2011): 216–254.

Köseoğlu, Mehmet. "Nüfus Kaynakları ve Sözlü Tarih Kaynaklarına Göre Balkan Savaşları Sırasında Kosova'dan Samsun'a Göçler." *History Studies* 5, no. 6 (2013): 17–40.

Koylu, Zafer, and Nihal Doğan. "Birinci Dünya Savaşı Sırasında Osmanlı Devleti'nde Sıtma Mücadelesi ve Bu Amaçla Yapılan Yasal Düzenlemeler." *Türkiye Parazitoloji Dergisi* 34 (2010): 209–215.

Krause, Franz. "Making Space along the Kemi River: A Fluvial Geography in Finnish Lapland." *Cultural Geographies* 42, no. 2 (2017): 279–294.

———. "Reclaiming Flow for a Lively Anthropology." *Suomen Antropologi: Journal of the Finnish Anthropological Society* 39, no. 2 (2014): 89–102.

———. "Rhythms of Wet and Dry: Temporalising the Land-Water Nexus." *Geoforum* (2017). https://doi.org/10.1016/j.geoforum.2017.12.001.

Kuijken, Eckhart. "A Short History of Waterbird Conservation." In *Waterbirds around the World*, edited by G. C. Boere, C. A. Galbraith, and D. A. Stroud, 52–59. Edinburgh: Stationery Office, 2006.

Kumerloeve, Hans. *Bibliographie der Säugetiere und Vögel der Türkei (rezente Fauna): Unter Berücksichtigung der Benachbarten Gebiete und mit Hinweisen auf Weiterführendes Schrifttum.* Bonn, Germany: Zoologisches Forschungsinstitut und Museum Alexander Koenig, 1986.

Kurtiç, Ekin. "Sediment in Reservoirs: A History of Dams and Forestry in Turkey." In *Transforming Socio-natures in Turkey: Landscapes, State and Environmental Movements*, edited by Onur İnal and Ethemcan Turhan, 90–111. New York: Routledge, 2020.

———. "Sedimented Encounters: Dams, Conservation, and Politics in Turkey." PhD diss., Harvard University, 2019.

Kurtiç, Ekin, and Zeynep Kadirbeyoğlu. "Problems and Prospects for Genuine Participation in Water Governance in Turkey." In *Contemporary Water Governance in the Global South: Scarcity, Marketization and Participation*, edited by Leila M. Harris, Jacqueline A. Goldin, and Christopher Sneddon, 199–215. London: Routledge, 2015.

"Kuşlar Kurtuldu Sıra Balıklarda." *Yeni Asır*, February 7, 2013. https://www.yeniasir.com.tr/izmir/2013/02/07/kuslar-kurtuldu-sira-baliklarda.

Larkin, Brian. "The Politics and Poetics of Infrastructure." *Annual Review of Anthropology* 42 (2013): 327–343.

———. *Signal and Noise: Media, Infrastructure, and Urban Culture in Nigeria.* Durham, NC: Duke University Press, 2008.

Latour, Bruno. *Reassembling the Social: An Introduction to Actor-Network-Theory.* Oxford: Oxford University Press, 2005.

———. "Why Has Critique Run out of Steam? From Matters of Fact to Matters of Concern." *Critical Inquiry* 30, no. 2 (2004): 225–248.

Latour, Bruno, and Steve Woolgar. *Laboratory Life: The Construction of Scientific Facts.* Princeton, NJ: Princeton University Press, 1979.

Lewis, Michael L. *Inventing Global Ecology: Tracking the Biodiversity Ideal in India, 1945–1997.* Athens: Ohio University Press, 2004.

Linton, Jamie. *What Is Water? The History of a Modern Abstraction.* Vancouver: University of British Columbia Press, 2010.

Lopes, Ricardo Jorge, Juan Antonio Gomez, Alessandro Andreotti, and Maura Andreoni. "Purple Swamphen or Gallinule (*Porphyrio porphyrio*) and Humans: Forgotten History of Past Interactions." *Society & Animals* 24, no. 6 (2016): 574–595.

Lorimer, Jamie. *Wildlife in the Anthropocene: Conservation after Nature.* Minneapolis: University of Minnesota Press, 2015.

Lowe, Celia. "Making the Monkey: How the Togean Macaque Went from 'New Form' to 'Endemic Species' in Indonesians' Conservation Biology." *Cultural Anthropology* 19, no. 4 (2004): 491–516.

———. *Wild Profusion: Biodiversity Conservation in an Indonesian Archipelago.* Princeton, NJ: Princeton University Press, 2006.

Magnin, Gernant, and Murat Yarar. *Important Bird Areas in Turkey.* Istanbul: Dogal Hayati Koruma Dernegi, 1997.

"Manisalı Üretici Üzümün Cenaze Namazını Kıldı." August 17, 2016. http://www. hurriyet.com.tr/manisali-uretici-uzumun-cenaze-namazini-kildi-40197996.

"Marmara Denizi'nin 2 Katı Sulak Alan Kurudu." *CNN Türk.* September 15, 2013. https://www.cnnturk.com/2013/turkiye/09/15/marmara.denizinin.2.kati.sulak. alan.kurudu/723480.0/index.html.

Martin, Aryn, Natasha Myers, and Ana Viseu. "The Politics of Care in Technoscience." *Social Studies of Science* 45, no. 5 (2015): 1–17.

Martin, Laura J. "Proving Grounds: Ecological Fieldwork in the Pacific and the Materialization of Ecosystems." *Environmental History* 23, no. 3 (2018): 567–592.

Martinez-Reyes, José E. *Moral Ecology of a Forest: The Nature Industry and Maya Post-conservation.* Tucson: University of Arizona Press, 2016.

Mathur, Anuradha, and Dilip da Cunha. *Mississippi Floods: Designing a Shifting Landscape.* New Haven, CT: Yale University Press, 2001.

Matthews, Geoffrey Vernon Townsend. *The Ramsar Convention on Wetlands: Its History and Development.* Gland, Switzerland: Ramsar Convention Bureau, 1993.

Mattingly, Cheryl, and Jason Throop. "The Anthropology of Ethics and Morality." *Annual Review of Anthropology* 47, no. 1 (2018): 475–492.

Mavhunga, Clapperton Chakanetsa. *Transient Workspaces: Technologies of Everyday Innovation in Zimbabwe.* Cambridge, MA: MIT Press, 2014.

McAfee, Kathleen. "Selling Nature to Save It? Biodiversity and Green Developmentalism." *Environment and Planning* 17, no. 2 (1999): 133–154.

McCarthy, Justin. *Muslims and Minorities: The Population of Ottoman Anatolia and the End of the Empire.* New York: New York University Press, 1983.

McGhie, Henry A. *Henry Dresser and Victorian Ornithology: Birds, Books and Business.* Manchester, UK: Manchester University Press, 2017.

McLean, Stuart. "Black Goo: Forceful Encounters with Matter in Europe's Muddy Margins." *Cultural Anthropology* 26, no. 4 (2011): 589–619.

Meichanetsidis, Vasileios. "The Genocide of the Greeks of the Ottoman Empire, 1913–1923: A Comprehensive Overview." *Genocide Studies International* 9, no. 1 (2015): 104–173.

Meiton, Fredrik. *Electrical Palestine: Capital and Technology from Empire to Nation.* Oakland: University of California Press, 2019.

Mengüllüoğlu, Deniz, Kazım Aktan, and Osman Erdem. *Göksu Deltası Özel Çevre*

Koruma Bölgesi Saz Horozu (Porphyrio Porphyrio) Koruma ve Gözleme Projesi Kesin Raporu. Ankara: Kuş Araştırmaları Derneği, 2008.

Meyer, William B. "When Dismal Swamps Became Priceless Wetlands." *American Heritage* 43, no. 3 (June 1994). https://www.americanheritage.com/when-dismal-swamps-became-priceless-wetlands.

Mikhail, Alan. "A Dog-Eat-Dog Empire: Violence and Affection on the Streets of Ottoman Cairo." *Comparative Studies of South Asia, Africa and the Middle East* 35, no. 1 (2015): 76–95.

———. *Nature and Empire in Ottoman Egypt: An Environmental History.* Cambridge: Cambridge University Press, 2011.

———. "Unleashing the Beast: Animals, Energy, and the Economy of Labor in Ottoman Egypt." *American Historical Review* 118, no. 2 (2013): 317–348.

Millar, Kathleen M. *Reclaiming the Discarded: Life and Labor on Rio's Garbage Dump.* Durham, NC: Duke University Press, 2018.

Milton, Giles. *Paradise Lost: Smyrna, 1922.* New York: Basic Books, 2008.

Mitchell, Timothy. "Can the Mosquito Speak?" In *Rule of Experts: Egypt, Techno-politics, Modernity,* by Timothy Mitchell, 19–53. Berkeley: University of California Press, 2002.

Mol, Annemarie. *The Logic of Care: Health and the Problem of Patient Choice.* New York: Routledge, 2008.

Moore, Donald S. *Suffering for Territory: Race, Place, and Power in Zimbabwe.* Durham, NC: Duke University Press, 2005.

Moore-Colyer, R. J. "Feathered Women and Persecuted Birds: The Struggle against the Plumage Trade, c. 1860–1922." *Rural History* 11, no. 1 (2000): 57–73.

Morita, Atsuro. "Infrastructuring Amphibious Space: The Interplay of Aquatic and Terrestrial Infrastructures in the Chao Phraya Delta in Thailand." *Science as Culture* 25, no. 1 (2016): 117–140.

———. "Multispecies Infrastructure: Infrastructural Inversion and Involutionary Entanglements in the Chao Phraya Delta, Thailand." *Ethnos* 82, no. 4 (2017): 738–757.

Moser, M., R. C. Prentice, and J. Van Vessem, eds. *Waterfowl and Wetland Conservation in the 1990s, a Global Perspective: Proceedings of an IWRB Symposium, St. Petersburg Beach, Florida, USA, 12–19 November 1992.* IWRB Special Publications. Slimbridge, UK: International Waterfowl and Wetlands Research Bureau, 1993.

Muehlebach, Andrea. *The Moral Neoliberal: Welfare and Citizenship in Italy.* Chicago: University of Chicago Press, 2012.

Mullin, Molly. "Mirrors and Windows: Sociocultural Perspectives on Human-Animal Relationships." *Annual Review of Anthropology* 28 (1999): 201–224.

Mutlu, Burcu. "Transnational Biopolitics and Family-Making in Secrecy: An Ethnography of Reproductive Travel from Turkey to Northern Cyprus." PhD diss., Massachusetts Institute of Technology, 2019.

Nadasdy, Paul. "The Gift in the Animal: The Ontology of Hunting and Human–Animal Sociality." *American Ethnologist* 34, no. 1 (2007): 25–43.

Navid, Daniel. "Letter to M. Paszkowski, Re. Accession of Turkey and Acceptance of

Amendments to Articles 7 and 7. 16 August 1994." Ramsar Office Archives, Gland, Switzerland.

Neumann, Roderick P. *Imposing Wilderness: Struggles over Livelihood and Nature Preservation in Africa.* Berkeley: University of California Press, 1998.

———. "Political Ecology of Wildlife Conservation in the Mt. Meru Area of Northeast Tanzania." *Land Degradation & Rehabilitation* 3 (1992): 85–98.

Neyzi, Leyla. *Ben Kimim? Türkiye'de Sözlü Tarih, Kimlik ve Öznellik.* Istanbul: İletişim Yayınları, 2009.

———. "Remembering Smyrna/Izmir: Shared History, Shared Trauma." *History and Memory: Studies in Representation of the Past* 2 (2008): 106–127.

Niethammer, Günther. "Die Vogelwelt von Auschwitz." *Annalen des Naturhistorischen Museums in Wien* 52 (1941): 164–199.

Nowak, Eugeniusz. *Biologists in the Age of Totalitarianism: Personal Reminiscences of Ornithologists and Other Naturalists.* Newcastle upon Tyne, UK: Cambridge Scholars Publishing, 2018.

Nucho, Joanne Randa. *Everyday Sectarianism in Urban Lebanon: Infrastructures, Public Services, and Power.* Princeton, NJ: Princeton University Press, 2016.

O'Connor, Terence P. "Working at Relationships: Another Look at Animal Domestication." *Antiquity* 71, no. 271 (1997): 149–156.

Öktem, Kerem. "The Nation's Imprint: Demographic Engineering and the Change of Toponymes in Republican Turkey." *European Journal of Turkish Studies* 7 (2008). https://doi.org/10.4000/ejts.2243.

Olney, Peter James Stephen. *Project Mar: The Conservation and Management of Temperate Marshes, Bogs and Other Wetlands.* Vol. 2, *List of European and North African Wetlands of International Importance.* IUCN Publications New Series. Morges, Switzerland: IUCN, International Waterfowl Research Bureau, ICBP (International Council for Bird Preservation), 1965.

Önder, Sylvia Wing. *We Have No Microbes Here: Healing Practices in a Turkish Black Sea Village.* Durham, NC: Carolina Academic Press, 2007.

Orbay, Faith, dir. *Wetlands, the Source of Life (Hayatın Kaynağı Sulak Alanlar).* Documentary film, 2000.

Orlove, Ben, and Steven C. Caton. "Water Sustainability: Anthropological Approaches and Prospects." *Annual Review of Anthropology* 39 (2010): 401–415.

"Orman Köylülerinin Kalkınmalarının Desteklenmesi ve Hazine Adına Orman Sınırları Diflina Çikarılan Yerlerin Değerlendirilmesiyle Hazineye Ait Tarım Arazilerinin Satışlı Hakkında Kanun." *Resmî Gazete* 28275 (April 26, 2012). http://www.resmigazete.gov.tr/eskiler/2012/04/20120426-1.htm.

Özesmi, Uygar. "Conservation Strategies for Sustainable Resource Use in the Kızılırmak Delta in Turkey." PhD diss., University of Minnesota, 1999.

———. "The Ecological Economics of Harvesting Sharp-Pointed Rush (*Juncus acutus*) in the Kizilirmak Delta, Turkey." *Human Ecology* 4 (2003): 645–655.

Özesmi, Uygar, Mehmet Somuncu, and Harun Tunçel. "Sultan Sazlığı Ekosistemi." In

Ankara Üniversitesi Türkiye Coğrafyası Araştırma ve Uygulama Merkezi Türkiye Coğrafyası Dergisi, 275–288. Ankara: Ankara Üniversitesi Basımevi, 1993.

Özgen, Can, and Beyhan Taş. "Ramsar Alanı İçinde Yer Alan Cernek Gölü Ve Sulak Alaninin (Kızılırmak Deltası, Samsun) Ekolojiv ve Sosyo-Ekonomik Önemi." *Tubav Bilim Dergisi* 2 (2012): 1–11.

Özkan, Hande. "Remembering Zingal: State, Citizens, and Forests in Turkey." *International Journal of Middle East Studies* 50, no. 3 (2018): 493–511.

Özkırımlı, Umut. *The Making of a Protest Movement in Turkey*. New York: Palgrave Macmillan, 2014.

Özkırlı, Burak Sait. "Flamingolar Çamura Gömülmesin." *Atlas Dergisi*, January 7, 2013. https://www.atlasdergisi.com/kesfet/doga-cografya-haberleri/flamingolar-camura-gomulmesin.html.

Özyürek, Esra, ed. *The Politics of Public Memory in Turkey*. Modern Intellectual and Political History of the Middle East. Syracuse, NY: Syracuse University Press, 2007.

Palmer, Theodore Sherman. *Legislation for the Protection of Birds Other Than Game Birds*. Washington, DC: Government Printing Office, 1902.

Pandian, Anand. "Pastoral Power in the Postcolony: On the Biopolitics of the Criminal Animal in South India." *Cultural Anthropology* 23, no. 1 (2008): 85–117.

Parla, Ayşe. "Critique without a Politics of Home?" In *A Time for Critique*, edited by Didier Fassin and Bernard Harcourt, 52–70. New York: Columbia University Press, 2019.

Parvin, Manoucher, and Mukerrem Hic. "Land Reform versus Agricultural Reform: Turkish Miracle or Catastrophe Delayed?" *International Journal of Middle East Studies* 16, no. 2 (1984): 207–232.

Paxson, Heather. *The Life of Cheese: Crafting Food and Value in America*. Berkeley: University of California Press, 2013.

———. "Post-Pasteurian Cultures: The Microbiopolitics of Artisanal Raw Milk Cheese in the United States." *Cultural Anthropology* 23, no. 1 (2008): 15–47.

Paxson, Heather, and Stefan Helmreich. "The Perils and Promises of Microbial Abundance: Novel Natures and Model Ecosystems, from Artisanal Cheese to Alien Seas." *Social Studies of Science* 44, no. 2 (2013): 165–193.

Pelkmans, Mathijs. "The Social Life of Empty Buildings: Imagining the Transition in Post-Soviet Ajaria." *Focaal: European Journal of Global and Historical Anthropology* 41 (2003): 121–135.

Peluso, Nancy Lee. *Rich Forests, Poor People: Resource Control and Resistance in Java*. Berkeley: University of California Press, 1994.

Pigg, Stacy Leigh. "Inventing Social Categories through Place: Social Representations and Development in Nepal." *Comparative Studies in Society and History* 34, no. 3 (1992): 491–513.

Polanyi, Karl. *The Great Transformation*. Boston: Beacon Press, 1957.

Popov, Anton. *Culture, Ethnicity and Migration after Communism: The Pontic Greeks*. London: Routledge, 2016.

Powell, Miles A. *Vanishing America: Species Extinction, Racial Peril, and the Origins of Conservation.* Cambridge, MA: Harvard University Press, 2016.

Pritchard, Sara B. *Confluence: The Nature of Technology and the Remaking of the Rhône.* Cambridge, MA: Harvard University Press, 2011.

Project Mar: The Conservation and Management of Temperate Marshes, Bogs, and Other Wetlands, First Volume. Proceedings of the MAR Conference Organized by IUCN, ICBP, and IWRB at Les Saints-Marie-de-La-Mer November 12–16 1992. IUCN Publications New Series. Morges, Switzerland: IUCN, International Waterfowl Research Bureau ICBP (International Council for Bird Preservation), 1964.

Prott, Volker. *The Politics of Self-Determination: Remaking Territories and National Identities in Europe, 1917–1923.* Oxford: Oxford University Press, 2016.

Quataert, Donald, ed. *Consumption Studies and the History of the Ottoman Empire, 1550–1922: An Introduction.* New York: State University of New York Press, 2000.

——. *Social Disintegration and Popular Resistance in the Ottoman Empire, 1881–1908: Reactions to European Economic Penetration.* New York: New York University Press, 1983.

Raffles, Hugh. *In Amazonia: A Natural History.* Princeton, NJ: Princeton University Press, 2002.

Ramsar Convention Secretariat. *Addressing Change in Wetland Ecological Character: Addressing Change in the Ecological Character of Ramsar Sites and Other Wetlands.* Ramsar Handbooks for the Wise Use of Wetlands, 4th ed., vol. 19. Gland, Switzerland: Ramsar Convention Secretariat, 2010.

Rangan, Haripriya. *Of Myths and Movements: Rewriting Chipko into Himalayan History.* London: Verso, 2000.

Rees, Lucy. *Horses in Company.* Ramsbury, Marlborough, UK: J. A. Allen, 2017.

Reisman, Arnold. *Turkey's Modernization: Refugees from Nazism and Atatürk's Vision.* Washington, DC: New Academic Publication, 2006.

Republic of Turkey Ministry of Environment. *The Kızılırmak Delta.* Ankara: Republic of Turkey Ministry of Environment General Directorate of Environmental Protection Wetland Section, 1998.

Richardson, Tanya. "The Terrestrialization of Amphibious Life in a Danube Delta 'Town on Water.'" *Suomen Antropologi: Journal of the Finnish Anthropological Society* 43, no. 2 (2018): 3–29.

——. "Where the Water Sheds: Disputed Deposits at the Ends of the Danube." In *The Poetics and Politics of the Danube River,* edited by Marjieta Bozovic, 308–337. Brookline, MA: Academic Studies Press, 2016.

Richardson, Tanya, and Gisa Weszkalnys. "Introduction: Resource Materialities." *Anthropological Quarterly* 87, no. 1 (2014): 5–30.

Riles, Annelise. "The Anti-network: Private Global Governance, Legal Knowledge, and the Legitimacy of the State." *American Journal of Comparative Law* 56, no. 3 (2008): 605–630.

Ritvo, Harriet. *The Animal Estate: The English and Other Creatures in the Victorian Age.* Cambridge, MA: Harvard University Press, 1989.

———. *The Dawn of Green: Manchester, Thirlmere and Modern Environmentalism.* Chicago: University of Chicago Press, 2009.

———. *The Platypus and the Mermaid: And Other Figments of the Classifying Imagination.* Cambridge, MA: Harvard University Press, 1998.

Rizvi, Mubbashir. "The Moral Ecology of Colonial Infrastructure and the Vicissitudes of Land Rights in Rural Pakistan." *History and Anthropology* 28, no. 3 (2017): 308–325.

Robbins, Paul. *Political Ecology: A Critical Introduction.* Chichester, West Sussex, UK: Wiley-Blackwell, 2012.

Russi, Daniela, Patrick ten Brink, Andrew Farmer, Tomas Badura, David Coates, Johannes Förster, Ritesh Kumar, and Nick Davidson. *The Economics of Ecosystems and Biodiversity (TEEB) for Water and Wetlands.* Gland, Switzerland: Convention on Biological Diversity, 2012.

Ruzicka, Stephen. *Trouble in the West: Egypt and the Persian Empire, 525–332 BC.* Oxford: Oxford University Press, 2012.

Sahakyan, Lusine. *Turkification of the Toponyms in the Ottoman Empire and the Republic of Turkey.* Montreal, Canada: Arod Books, 2010.

Saltan, Sevket, and Cahit Okar. *A Summary of the Salt Industry in Turkey Çamaltı Saltworks.* Rome: United Nationals Industrial Development Organization, 1963.

Sanal, Aslıhan. *New Organs within Us: Transplants and the Moral Economy.* Durham, NC: Duke University Press, 2011.

Scaramelli, Caterina. "The Delta Is Dead: Moral Ecologies of Infrastructure in Turkey." *Cultural Anthropology* 34, no. 3 (2019): 388–416.

———. "Fish, Flows, and Desire in the Delta." *Anthropology News* 59, no. 2 (2018): 3–5.

———. "Making Sense of Water Quality." *Worldviews* 17, no. 2 (2013): 150–160.

———. "Swamps into Wetlands: Making Livable Nature in Turkey." PhD diss., Massachusetts Institute of Technology, 2016.

———. "The Wetland Is Disappearing: Conservation and Care on Turkey's Kızılırmak Delta." *International Journal of Middle East Studies* 50, no. 3 (2018): 405–425.

Schiappa, Edward. "Towards a Pragmatic Approach to Definition: Wetlands and the Politics of Meaning." In *Environmental Pragmatism*, edited by Andrew Light and Eric Katz, 209–230. New York: Routledge, 1996.

Schwenkel, Christina. "Spectacular Infrastructure and Its Breakdown in Socialist Vietnam." *American Ethnologist* 42, no. 3 (2015): 520–534.

Scott, James C. *The Moral Economy of the Peasant: Rebellion and Subsistence in Southeast Asia.* New Haven, CT: Yale University Press, 1977.

Şekercioğlu, Çağan, Sean Anderson, Erol Akçay, Raşit Bilgin, Emre Can Özgün, Semiz Gürkan, Çağatay Tavşanoğlu, et al. "Turkey's Globally Important Biodiversity in Crises." *Biological Conservation* 144, no. 12 (2011): 2752–2769.

Selçuk, Zehra. "Anadolu Mandalarının Doğal Yaşam Alanı 'Kızılırmak Deltası.'" *Kafkas Universitesi Veteriner Fakultesi Dergisi* 18, no. 1 (2012): 167.

Shaw, Samuel P., and Gordon Fredine. *Wetlands of the United States.* Washington, DC: US Government Printing Office, 1956.

Sıkı, Mehmet. "Izmir Kuş Cennetinin Tarihçesi." *Tabiat ve Insan* 27, no. 1 (1994): 6–10.

———. "Izmir Kuşcenneti'nin Dünü, Bügünü, ve Yarını." In *II Kıyı Sorunlar ve Çevre Sempozyumu, November 14–16*, edited by Ayşe Günbey Şerifoğlu, 748–756. Kuşadası, Turkey: Belediye Yayınları, 1997.

Sıkı, Mehmet, and Ibrahim Baran. "Çamaltı Tuzlası'ndaki 'Kuş Cenneti.'" *Bilim ve Teknik* (April 1984): 4–5.

Simone, AbdoulMaliq. "People as Infrastructure: Intersecting Fragments in Johannesburg." *Public Culture* 16, no. 3 (2004): 407–429.

Sirman, Nükhet. "State, Village, and Gender in Western Turkey." In *Turkish State, Turkish Society*, edited by Nükhet Sirman and Andrew Finkel, 283–310. New York: Routledge, 1990.

"Sıtma Mücâdelesi Kânûnu." *Resmî Gazete* 384 (May 29, 1926). https://www.resmigazete.gov.tr/arsiv/384.pdf.

Sivaramakrishnan, Kalyanakrishnan. "Ethics of Nature in Indian Environmental History." *Modern Asian Studies* 49, no. 4 (2015): 1261–1310.

———. *Modern Forests: Statemaking and Environmental Change in Colonial Eastern India*. Stanford, CA: Stanford University Press, 1999.

Snowden, Frank. *The Conquest of Malaria: Italy, 1900–1962*. New Haven, CT: Yale University Press, 2008.

Sönmez, Ipek, and Ortaç Onmuş. *Gediz Deltası Yönetim Planı Sosyo-Ekonomic Analiz Raporu*. Izmir: Izmir Kuş Cennetini Koruma ve Geliştirme Birliği, Ege Doğal Yaşamı Koruma Derneği, 2006.

Stamatopoulou-Robbins, Sophia. *Waste Siege: The Life of Infrastructure in Palestine*. Stanford, CA: Stanford University Press, 2019.

Staples, Amy. *The Birth of Development: How the World Bank, Food and Agriculture Organization, and World Health Organization Changed the World, 1945–1965*. Kent, OH: Kent State University Press, 2006.

Star, Susan Leigh. "The Ethnography of Infrastructure." *American Behavioral Scientist* 43, no. 3 (1999): 377–391.

Star, Susan Leigh, and James R. Griesemer. "Institutional Ecology, 'Translations' and Boundary Objects: Amateurs and Professionals in Berkeley's Museum of Vertebrate Zoology, 1907–39." *Social Studies of Science* 19, no. 3 (1989): 387–420.

Stoetzer, Bettina. "Ruderal Ecologies: Rethinking Nature, Migration, and the Urban Landscape in Berlin." *Cultural Anthropology* 33, no. 2 (2018): 295–323.

Strabo of Amaseia. *Complete Works of Strabo*. Hastings, West Sussex, UK: Delphi Classics, 2016.

Strang, Veronica. *The Meaning of Water*. Oxford, UK: Berg Publishers, 2004.

Subramaniam, Banu. "The Aliens Have Landed! Reflections on the Rhetoric of Biological Invasions." *Meridian* 2, no. 1 (2001): 26–40.

Sulak Alanlar Şubesi Müdürlüğü. *Gediz Deltası Sulakalan Yönetim Planı*. Ankara: Doğa Koruma ve Milli Parklar Genel Müdürlüğü, 2007.

"Sulak Alanların Korunması Yönetmenliği." *Resmî Gazete* 28962 (April 4, 2014). https://www.resmigazete.gov.tr/eskiler/2014/04/20140404-11.htm.

"Suriyeli Çocuklardan Temsili Cenaze Töreni." *Hürriyet*, June 24, 2014. http://www. hurriyet.com.tr/dunya/suriyeli-cocuklardan-temsili-cenaze-toreni-26665555.

Takacs, David. *The Idea of Biodiversity: Philosophies of Paradise*. Baltimore: John Hopkins University Press, 1996.

Tambar, Kadir. "Brotherhood in Dispossession: State Violence and the Ethics of Expectation in Turkey." *Cultural Anthropology* 31, no. 1 (2016): 30–55.

Tambiah, Stanley Jeyaraja. "Animals Are Good to Think and Good to Prohibit." *Ethnology* 8, no. 4 (1969): 423–459.

"Tansu Gürpinar'la Söyleşi." *Fotoritim*, February 1, 2009. https://web.archive.org/web/20171001135053/http://www.arsivfotoritim.com/yazi/tansu-gurpinar-ile-soylesi/.

Technical Assistance for the Integrated Management of Wetlands in the Gediz Delta 2009–2014. Gland, Switzerland: Tour du Valat, 2014.

Thompson, E. P. "The Moral Economy of the English Crowd in the Eighteenth Century." *Past & Present* 50 (1971): 76–136.

Tronto, Joan C. *Moral Boundaries: A Political Argument for an Ethic of Care*. London: Routledge, 1993.

Tsing, Anna Lowenhaupt. *The Mushroom at the End of the World: On the Possibility of Life in Capitalist Ruins*. Princeton, NJ: Princeton University Press, 2015.

Turam, Berna. *Between Islam and the State: The Politics of Engagement*. Stanford, CA: Stanford University Press, 2007.

Turner, Michael. "Thomas Robson 1812–1884: The Forgotten Bird Man." In *Northumbrian Naturalists: Transactions of the Natural History Society*, edited by Chris Redfern, 75:42–52. Newcastle upon Tyne, UK: Natural History Society of Northumbria, 2013.

UNDP GEF. *Conservation of Biodiversity through Improvement of Water Buffalo Farming in Kızılırmak Delta*. GEF Small Grants Programme, 2009. https://sgp.undp.org/index.php?option=com_sgpprojects&view=projectdetail&id=14176&Itemid=272.

UNESCO. "The Criteria for Selection." Accessed April 7, 2020. https://whc.unesco.org/en/criteria/.

Üngör, Uğur Ümit. *The Making of Modern Turkey: Nation and State in Eastern Anatolia, 1913–1950*. Oxford: Oxford University Press, 2011.

Van Dooren, Thom. *Flight Ways: Life and Loss at the Edge of Extinction*. New York: Columbia University Press, 2014.

Vitebsky, Piers. *The Reindeer People: Living with Animals and Spirits in Siberia*. Boston: Houghton Mifflin Harcourt, 2006.

Von Schnitzler, Antina. "Traveling Technologies: Infrastructure, Ethical Regimes, and the Materiality of Politics in South Africa." *Cultural Anthropology* 28, no. 4 (2013): 670–693.

Voulvouli, Aimilia. *From Environmentalism to Transenvironmentalism: The Ethnography of an Urban Protest in Modern Istanbul*. Bern, Switzerland: Peter Lang, 2009.

Wahby, Ali. "Les oiseaux de la region de Stamboul et ses environs." *Bulletin de la Société Zoologique de Genève* 9, no. 1 (1929): 171–175.

Walley, Christine J. *Rough Waters: Nature and Development in an East African Marine Park*. Princeton, NJ: Princeton University Press, 2004.

Water for People, Water for Life: The United Nations World Water Development Report. Barcelona: UNESCO and Berghahn Books, 2003.

Watson, George E. "Aegean Bird Notes 1: Descriptions of New Subspecies from Turkey." *Postilla, Yale Peabody Museum of Natural History* 52 (June 28, 1963): 1–15.

West, Paige. *Conservation Is Our Government Now: The Politics of Ecology in Papua New Guinea*. Durham, NC: Duke University Press, 2006.

West, Paige, James Igoe, and Dan Brockington. "Parks and Peoples: The Social Impact of Protected Areas." *Annual Review of Anthropology* 35 (2006): 251–277.

White, Jenny. *Muslim Nationalism and the New Turks*. Princeton Studies in Muslim Politics. Princeton, NJ: Princeton University Press, 2013.

White, Richard. *The Organic Machine: The Remaking of the Columbia River*. New York: Hill and Wang, 1996.

White, Sam. *The Climate of Rebellion in the Early Modern Ottoman Empire*. Cambridge: Cambridge University Press, 2011.

———. "Rethinking Disease in Ottoman History." *International Journal of Middle East Studies* 42, no. 4 (2010): 549–567.

Williams, Raymond. "Dominant, Residual, and Emergent." In *Marxism and Literature*, by Raymond Williams, 121–127. New York: Oxford University Press, 1977.

Wilson, Robert M. *Seeking Refuge: Birds and Landscapes of the Pacific Flyway*. Seattle: University of Washington Press, 2011.

Yates, Julian S., Leila M. Harris, and Nicole J. Wilson. "Multiple Ontologies of Water: Politics, Conflict and Implications for Governance." *Environment and Planning D: Society and Space* 35, no. 5 (2017): 797–815.

Yates-Doerr, Emily. "Does Meat Come from Animals? A Multispecies Approach to Classification and Belonging in Highland Guatemala." *American Ethnologist* 42, no. 2 (2015): 309–323.

Yazıcı, Meryem, and Baha Büyükışık. "Homa Dalyanı (İzmir Körfezi, Ege Denizi)'nda Sekonder Prodüktivitenin Araştırılması." *Su Ürünleri Dergisi* 24, no. 3–4 (2007): 267–272.

Yeh, Emily. "From Wasteland to Wetland? Nature and Nation in China's Tibet." *Environmental History* 14, no. 1 (2009): 103–137.

Yeniyürt, Can, Serhan Çağırankaya, Lise Yıldıray, and Yusuf Ceran. *Kızılırmak Deltası Sulak Alan Yönetim Planı 2008–2012*. Ankara: Çevre Bakanlığı, Doğa Koruma ve Milli Parklar Genel Müdürlüğü, Doğa Koruma Dairesi Başkanlığı, Sulak Alan Şube Müdürlüğü, 2008.

Yıldırım, Mine. "Hayvan Tecritinin Dışı ve Ötesi: İstanbul'da Sürgün, Yıkım ve Şiddet Coğrafyaları." *Beyond.Istanbul* 4 (2019): 87–99.

Yılmaz, Cevdet. "Altınkaya Barajı'nın Vezirköprü'ye Etikleri." *Eskişehir Osmangazi Üniversitesi Sosyal Bilimler Dergisi* 1, no. 1 (2000): 137–158.

———. *Bafra Ovası'nın Beşeri ve İktisadi Coğrafyası*. Ankara: Gündüz Eğitim Yayınları, 2002.

Yılmaz, Orhan, Mehmet Ertuğrul, and Richard Trevor Wilson. "Domestic Livestock Resources of Turkey." *Tropical Animal Health and Production* 44, no. 4 (2012): 707–714.

Zandi-Sayek, Sibel. *Ottoman Izmir: The Rise of a Cosmopolitan Port, 1840–1880*. Minneapolis: University of Minnesota Press, 2012.

Zengin, Aslı. "The Afterlife of Gender: Sovereignty, Intimacy, and Muslim Funerals of Transgender People in Turkey." *Cultural Anthropology* 34, no. 1 (2019): 78–102.

Zeybek, Sezai Ozan. "Türkiye'de Endüstriel Hayvancılığın Seyri." *Toplum ve Bilim* 138 (2016): 7–25.

Zurcher, Erik J. *Turkey: A Modern History*. London: I. B. Tauris, 1997.

Index

CPSIA information can be obtained
at www.ICGtesting.com
Printed in the USA
JSHW032258300621
16479JS00001B/22